T0076251

METHODS IN MOLECULAR BIOLOGY™

Series Editor
John M. Walker
School of Life Sciences
University of Hertfordshire
Hatfield, Hertfordshire, AL10 9AB, UK

For further volumes:
http://www.springer.com/series/7651

Germline Development

Methods and Protocols

Edited by

Wai-Yee Chan

School of Biomedical Sciences, The Chinese University of Hong Kong, Shatin, N.T., Hong Kong SAR, China

Le Ann Blomberg

ARS–USDA, Animal Biosciences & Biotechnology Lab, Beltsville, MD, USA

Humana Press

Editors
Wai-Yee Chan, Ph.D.
School of Biomedical Sciences
The Chinese University of Hong Kong
Shatin, N.T., Hong Kong SAR, China
chanwy@cuhk.edu.hk

Le Ann Blomberg, Ph.D.
ARS-USDA
Animal Biosciences & Biotechnology Lab
Beltsville, MD, USA
leann.blomberg@ars.usda.gov

ISSN 1064-3745   e-ISSN 1940-6029
ISBN 978-1-61779-435-3   e-ISBN 978-1-61779-436-0
DOI 10.1007/978-1-61779-436-0
Springer New York Dordrecht Heidelberg London

Library of Congress Control Number: 2011941357

© Springer Science+Business Media, LLC 2012
All rights reserved. This work may not be translated or copied in whole or in part without the written permission of the publisher (Humana Press, c/o Springer Science+Business Media, LLC, 233 Spring Street, New York, NY 10013, USA), except for brief excerpts in connection with reviews or scholarly analysis. Use in connection with any form of information storage and retrieval, electronic adaptation, computer software, or by similar or dissimilar methodology now known or hereafter developed is forbidden.
The use in this publication of trade names, trademarks, service marks, and similar terms, even if they are not identified as such, is not to be taken as an expression of opinion as to whether or not they are subject to proprietary rights.

Printed on acid-free paper

Humana Press is part of Springer Science+Business Media (www.springer.com)

Preface

Proper development of the germ cells (i.e., germline) is essential for the propagation of mammalian species. In spite of recent advances in the biological sciences, our knowledge of the mechanisms that regulate germline development is still rather limited. This volume aims to provide a comprehensive collection of techniques that can be applied to the study of both the male and female germline.

The germline is unique in mammals as it is the only cell lineage that undergoes mitosis, meiosis, and differentiation. Furthermore, the lineage comprises an array of developmentally distinct stages from the germinal stem cell to the terminally differentiated sperm or oocyte. Fortunately, cells at unique stages of development can be isolated as relatively pure populations. This feature makes these cells amenable to in-depth genomic and mechanistic studies. Knowledge of the biology of cells of either the male or female lineage will provide us with a better appreciation of the cellular mechanisms modulating the germline developmental processes. In addition, understanding how a germinal stem cell maintains stemness versus the induction of its differentiation will provide us insight on how to direct the pathway of differentiation of stem cells in regenerative studies.

Past studies of germ cells have been largely limited to histological and cell biology aspects of these cells. Only limited information has been available on the molecular genetics and genomics of cells of the germline. With advances in technology in recent years, we are now at a stage where we can acquire a much better understanding of the genetic and genomic regulatory mechanisms of the developing germ cells. This volume aims at bringing the field of germline developmental studies together such that readers can find, in a single volume, techniques that allow the isolation of germ cells at different stages of development, the study of the genomics and epigenomics of germ cells, and an examination of the role of single genes in the development potential and function of germ cells.

Elucidation of the biology of germline development as well as germinal stem cell maintenance or differentiation would be of great impact to both the biomedical as well as agricultural communities. For example, improvements in assisted reproduction (in humans) and cloning (in animals) technologies are dependent on advancing current in vitro systems towards more efficient oocyte development. Similarly, the development of methodologies to maintain germinal stem cell systems for male or female instances of spontaneous or drug/toxin-induced infertility in humans or, in the case of agricultural animals, for alternative modes of cryopreservation to maintain germplasm or to improve transgenic technologies for the enhancement of production traits and the generation of animal models with nutraceutical worth, are all important.

The male and the female germline differ slightly in their developmental program. Consequently, the techniques used for isolating and studying male and female germ cells are slightly different. Therefore, this volume groups the different technologies under two sections, namely, male germline and female germline. It is our hope that this volume will provide a comprehensive text to individuals, in particular, researchers in reproductive and developmental biology, who are interested in studying cells of the germ lineage.

Shatin, N.T., Hong Kong   *Wai-Yee Chan*
Beltsville, MD, USA   *Le Ann Blomberg*

Contents

PART I MALE GERMLINE

Contributors

ALEXANDER I. AGOULNIK • *Herbert Wertheim College of Medicine, Florida International University, Miami, FL, USA*

RICHARD A. ANDERSON • *MRC Centre for Reproductive Health, The Queen's Medical Research Institute, University of Edinburgh, Edinburgh, UK*

ERIC ANTONIOU • *The Jackson Laboratory, Bar Harbor, ME, USA*

VANESSA BAXENDALE • *Eunice Kennedy Shriver National Institute of Child Health and Human Development, National Institutes of Health, Bethesda, MD, USA*

CATHERINE BOUCHERON • *Eunice Kennedy Shriver National Institute of Child Health and Human Development, National Institutes of Health, Bethesda, MD, USA*

PRABUDDHA CHAKRABORTY • *Department of Cellular and Integrative Physiology, University of Nebraska Medical Center, Omaha, NE, USA*

WAI-YEE CHAN • *School of Biomedical Sciences, The Chinese University of Hong Kong, Shatin, N.T., Hong Kong SAR, China*

HOI-HUNG CHEUNG • *Section on Clinical and Developmental Genomics, Eunice Kennedy Shriver National Institute of Child Health and Human Development, Bethesda, MD, USA*

ANDREW J. CHILDS • *MRC Centre for Reproductive Health, The Queen's Medical Research Institute, University of Edinburgh, Edinburgh, UK*

MARTINE CULTY • *The Research Institute of the McGill University Health Centre and Departments of Medicine and Pharmacology and Therapeutics, McGill University, Montreal, QC, Canada*

INA DOBRINSKI • *Department of Comparative Biology and Experimental Medicine, Faculty of Veterinary Medicine, University of Calgary, Calgary, AB, Canada*

MARTIN DYM • *Department of Biochemistry and Molecular and Cellular Biology, Georgetown University Medical Center, Washington, DC, USA*

LYDIA FERGUSON • *Herbert Wertheim College of Medicine, Florida International University, Miami, FL, USA*

THOMAS GARCIA • *Department of Comparative Biosciences, College of Veterinary Medicine, University of Illinois at Urbana-Champaign, Urbana, IL, USA*

ZUPING HE • *Clinical Stem Cell Research Center, Renji Hospital, Shanghai Jiao Tong University School of Medicine, Shanghai, China*

MARIE-CLAUDE HOFMANN • *Department of Comparative Biosciences, College of Veterinary Medicine, and Institute for Genomic Biology, University of Illinois at Urbana-Champaign, Urbana, IL, USA*

ZAOHUA HUANG • *Herbert Wertheim College of Medicine, Florida International University, Miami, FL, USA*

JIJI JIANG • *Department of Biochemistry and Molecular and Cellular Biology, Georgetown University Medical Center, Washington, DC, USA*

HISATO KOBAYASHI • *Department of BioScience, Tokyo University of Agriculture, Tokyo, Japan*

MARIA KOKKINAKI • *Department of Biochemistry and Molecular and Cellular Biology, Georgetown University Medical Center, Washington, DC, USA*

TOMOHIRO KONO • *Department of BioScience, Tokyo University of Agriculture, Tokyo, Japan*

REBECCA L. KRISHER • *University of Illinois at Urbana-Champaign, Urbana, IL, USA; National Foundation for Fertility Research, Lone Tree, CO, USA*

YUN-FAI CHRIS LAU • *Department of Medicine, VA Medical Center, University of California, San Francisco, CA, USA*

TIN-LAP LEE • *School of Biomedical Sciences, The Chinese University of Hong Kong, Shatin, Hong Kong, China; Section on Clinical and Developmental Genomics, Eunice Kennedy Shriver National Institute of Child Health and Human Development, Bethesda, MD, USA*

YUNMIN LI • *Department of Medicine, VA Medical Center, University of California, San Francisco, CA, USA*

CHANG LIU • *Department of Animal Sciences, University of Illinois at Urbana-Champaign, Urbana, IL, USA*

GURPREET MANKU • *The Research Institute of the McGill University Health Centre and Departments of Medicine and Pharmacology and Therapeutics, McGill University, Montreal, QC, Canada*

MONTY MAZER • *The Research Institute of the McGill University Health Centre and Departments of Medicine, McGill University, Montreal, QC, Canada*

ANINDIT MUKHERJEE • *Department of Cellular and Integrative Physiology, University of Nebraska Medical Center, Omaha, NE, USA*

MANAL OTHMAN • *Department of Comparative Biosciences, University of Illinois at Urbana-Champaign, Urbana, IL, USA*

MELISSA PACZKOWSKI • *Department of Comparative Biosciences, University of Illinois at Urbana-Champaign, Urbana, IL, USA*

ALAN LAP-YIN PANG • *Section on Clinical Genomics, Laboratory of Clinical and Developmental Genomics, Eunice Kennedy Shriver National Institute of Child Health and Human Development, National Institutes of Health, Bethesda, MD, USA*

OWEN M. RENNERT • *Section on Clinical and Developmental Genomics, Eunice Kennedy Shriver National Institute of Child Health and Human Development, Bethesda, MD, USA*

SHYAMAL K. ROY • *Departments of OB/GYN and Cellular and Integrative Physiology, University of Nebraska Medical Center, Omaha, NE, USA*

ROBERT TAFT • *The Jackson Laboratory, Bar Harbor, ME, USA*

TERUKO TAKETO • *Departments of Surgery, Biology, and Obstetrics and Gynecology, McGill University, Royal Victoria Hospital, Montreal, QC, Canada*

CHENG WANG • *Department of Obstetrics and Gynecology, University of Nebraska Medical Center, Omaha, NE, USA*

YAN WANG • *Department of Biochemistry and Molecular and Cellular Biology, Georgetown University, Washington, DC, USA*

AMY XIAO • *Laboratory of Clinical and Developmental Genomics, Eunice Kennedy Shriver National Institute of Child Health and Human Development, National Institutes of Health, Bethesda, MD, USA*

HUMPHREY HUNG-CHANG YAO • *Department of Comparative Biosciences, University of Illinois at Urbana-Champaign, Urbana, IL, USA; Laboratory of Reproductive and Developmental Toxicology, National Institute of Environmental Health Sciences (NIEHS/NIH), National Institutes of Health, Research Triangle Park, NC, USA*

YE YUAN • *University of Missouri, Columbia, MO, USA*

WENXIAN ZENG • *Northwest Agricultural and Forest University, Shanxi, China*

Part I

Male Germline

Chapter 1

Isolation of Fetal Gonads from Embryos of Timed-Pregnant Mice for Morphological and Molecular Studies

Yunmin Li, Teruko Taketo, and Yun-Fai Chris Lau

Abstract

Gonadal sex differentiation is an important developmental process, in which a bipotential primordial gonad undergoes two distinct pathways, i.e., testicular and ovarian differentiation, dependent on its genetic sex. Techniques of isolating fetal gonads at various developmental stages are valuable for studies on the molecular events involved in cell-fate determination, sex-specific somatic and germ-cell differentiation and structural organization. Here we describe various procedures for isolation of embryonic gonads at different developmental stages from embryos of timed-pregnant mice. The isolated fetal gonads can be used for a variety of studies, such as organ culture, gene and protein expression. As examples of applications, we describe the immunofluorescence detection of SOX9 expression in gonadal tissue sections and microRNAs profiling/expression in fetal gonads at a critical stage for sex determination.

Key words: Timed-pregnant mice, Fetal gonad, Mesonephros, Testis cords, Immunofluorescence, microRNAs profiling

1. Introduction

Gonad development is a complex organogenesis involving the sexual dimorphic differentiation of both somatic and germ cells (1, 2). The laboratory mouse has been a key model for research in gonadal differentiation, in which precise morphological and cellular changes during the developmental process(es) have been well documented and characterized. The mouse gonad emerges at the coelomic surface of mesonephros at 9.5 days post coitum (dpc) in a bipotential manner, being morphologically identical in XX and XY embryos. At 10.5–11.5 dpc, *Sry* begins to be expressed in the XY gonad (3) and initiates the differentiation of a subset of somatic cells into Sertoli cells by directly upregulating *Sox9* (4) and other immediate downstream

Wai-Yee Chan and Le Ann Blomberg (eds.), *Germline Development: Methods and Protocols*, Methods in Molecular Biology, vol. 825, DOI 10.1007/978-1-61779-436-0_1, © Springer Science+Business Media, LLC 2012

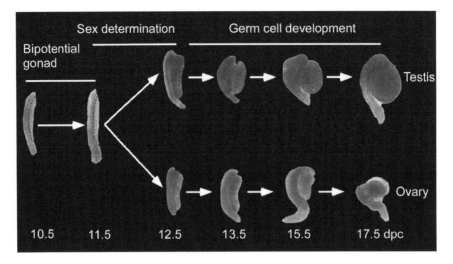

Fig. 1. Morphology of gonads isolated at different developmental stages. The testis cords are seen in the male gonads at 12.5 dpc and later but not at earlier stages.

testis differentiating genes. Upon Sertoli cell differentiation, the testis cords are formed, distinguishing morphologically the male gonads from female gonads (see Fig. 1). Ectopic expression of *Sry* in XX gonads induces testis development (5). Therefore, ovary development used to be thought as a default pathway. However, recent studies suggest that ovarian determining genes, such as *Rspo1* (6), *Wnt4* (7), and *FoxL2* (8) play important roles in ovary determination and differentiation in mammals.

Gonadogenesis also involves differentiation of primordial germ cells (PGCs). In the developing mouse embryo, PGCs migrate from the extraembryonic mesoderm into the urogenital ridge and populate the gonads between 10.0 and 11.5 dpc. Once inside the gonad, the germ cells continue to proliferate and go through several rounds of mitotic divisions in both sexes. Then the germ cells lose their motility and start to colonize in the gonad. Depending on the sex of gonadal somatic cells, germ cells in male and female gonads undergo two distinct developmental pathways. Male germ cells are arrested as prospermatogonia at G1/G0 from 13.5 dpc to a few days after birth (9). On the other hand, germ cells in the ovary enter the prophase of the first meiotic division and are arrested at the diplotene stage at birth.

Isolation of fetal gonads provides opportunity to study the sex differentiation of gonads and germ cells at morphological and molecular levels (3, 10, 11). Recombinant organ culture of gonads from wild-type mice (CD1) and mesonephroi from the transgenic mice ubquitously expressing β-galactosidase (ROSA26) has shown the importance of the migration of mesonephric cells into the XY but not XX gonad for testis cord formation (12). Microarray analysis of

Sf1-GFP-positive somatic cells isolated from gonads has made it possible to identify many upregulated and downregulated genes associated with sex differentiation (13, 14). In this chapter, we describe a protocol for the preparation of timed-pregnant mice, isolation of urogenital complexes from embryos, separation of the gonad from the mesonephros, and the method to sex the embryos younger than 12.5 dpc. Immunofluorescence of histological sections and miRNA (miRNAs) profiling of male and female gonads at 12.5 dpc will be illustrated as the examples of morphological and molecular studies on dissected fetal mouse gonads.

2. Materials

1. 70% ethanol.
2. Regular surgical forceps and scissors.
3. Dumont #5 Forceps, or any dissection forceps with straight fine tips.
4. Dumont #5/45 Forceps, or dissection forceps with fine tips angled at 45 degrees.
5. Micro Dissecting Scissors (ROBOZ RS-5880).
6. Micro Dissecting Knife-Needle, 4 1/2″, size 1 (Biomedical Research Instruments, Inc.).
7. Stereomicroscope.
8. Tissue culture dish (10-cm).
9. MIRASCORB Sponge.
10. 6–8 week old CD1 male and female mice (Charles River Laboratory).
11. Dulbecco's modified Eagle's medium (DMEM).
12. Fetal bovine serum (FBS).
13. Dissection medium: DMEM containing 10% FBS, 100 U/ml penicillin G and 100 μg/ml streptomycin, sterilized through 0.2-μm filter.
14. Phosphate-buffered saline (PBS).
15. 1.5-ml microcentrifuge tube.
16. 0.2-ml 8-strip PCR tubes.
17. DirectPCR lysis reagent (Viagen Biotech, Inc.).
18. Proteinase K.
19. RNAlater tissue stabilization solution (Ambion).
20. GoTaq DNA polymerase (Promega).
21. GeneAmp® PCR system 9700 (Applied Biosystems).

3. Methods

3.1. Preparation of Timed-Pregnant Females

Timed-pregnant mice can be purchased from commercial vendors, but since the time point is not always accurate, we prefer preparing timed-pregnant mice in our own laboratory.

1. Obtain 6–8 week old CD1 male and female mice from Charles River Laboratory, or other vendors.
2. Pair up female mice with male mice in cages just prior to the end of the daily light cycle.
3. Check the females for vaginal plugs early next morning. When a plug is found, the female is considered as pregnant at 0.5 dpc.
4. Transfer the plugged females into a new cage marked with the date on which a plug was detected and house them until dissection.

3.2. Isolation of Gonads from Embryos at 12.5 dpc or Later

Procedure for the isolation of gonads from 12.5 dpc embryos is shown as an example. This procedure can be used to isolate gonads from other gestational ages of embryos from timed-pregnant mice (Fig. 1). Dissection is usually carried out around noon on each day.

1. Sacrifice the timed-pregnant CD1 mice by cervical dislocation or other approved methods of the host Institutional Animal Care and Use Committee (IACUC).
2. Place the mouse on its back on absorbent paper and spray the mouse with 70% Ethanol thoroughly.
3. Lift the skin over the belly with forceps, make a lateral incision with regular surgical scissors, and pull the skin toward the head and the tail until the abdomen is completely exposed.
4. Open the abdominal cavity by lifting the abdominal muscle with forceps and cutting with scissors.
5. Grab the uterus with one set of forceps just above the cervix and cut the uterus cross the cervix with scissors. Then pull the uterus upward and cut beneath the oviduct.
6. Transfer the uterus into a 10-cm culture dish containing PBS.
7. Hold one end of the uterus with regular surgical forceps, insert one tip of a pair of Micro Dissecting Scissors into the antimesometrial wall of the uterus near the holding forceps, and cut the muscle by carefully pointing and sliding the scissors along the uterus. The embryos in amnionic membranes attached to the placenta will be exposed.
8. Separate each embryo from its placenta and surrounding membranes by carefully holding the placenta with forceps and cutting with Micro Dissecting Scissors (see Note 1).
9. Transfer the embryos to a new 10-cm dish and wash embryos with fresh PBS to remove the blood.

10. Place one layer of sponge in a 10-cm culture dish, add small amount of dissection medium to the dish (just enough to cover the bottom), and transfer the embryos onto the wet sponge (see Note 2).

11. Place the dish under a dissecting stereomicroscope, rotate the embryo on its side, hold the embryo with Dumont #5 forceps and cut off the anterior half of the embryo just below the forelimbs with Dumont #5/45 forceps (Fig. 2a, step 1). Then cut along the ventral midline of the posterior half of the embryo with Dumont #5/45 forceps, remove the liver and intestine (Fig. 2a, step 2), and cut off the hindlimbs and tail (Fig. 2a, step 3) (see Note 3).

12. Turn the embryo, and carefully remove the remaining intestines and liver. The urogenital complex must be visible on the dorsal wall of the embryo.

13. Hold the embryo with Dumont #5 forceps, cut out the urogenital complex with Dumont #5/45 forceps, and place it in a new 10-cm dish containing dissection medium.

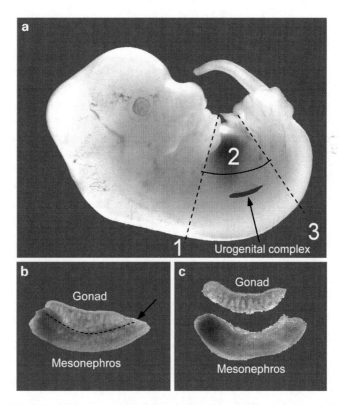

Fig. 2. Illustration of the procedure for the isolation of gonads from embryos at 12.5 dpc. (**a**) 12.5 dpc embryo, 1, 2, and 3 indicate the step of cutting, the position of urogenital complex is drawn on the embryo. (**b** and **c**) show the separation of gonad from mesonephros. The *dash line* in **b** shows the position for cutting.

14. Remove the connective tissues from the urogenital complex by holding with Dumont #5 forceps and cutting with Dumont #5/45 forceps.

15. Check the presence of testis cords in the dissected gonads under the dissecting stereomicroscope. At 12.5 dpc or later, the sex of the gonads can be morphologically distinguished by the presence of testicular cords in the testis.

16. Place the male and female urogenital complexes in separate 10-cm dishes containing dissection medium.

17. For immunohistochemistry, transfer the male and female urogenital complexes into the tubes containing a fixative (see Note 4).

18. For analyses of mRNA and protein, the gonads must be separated from the adjacent mesonephroi. Hold the mesonephros with Dumont #5 forceps and gently cut the connective tissue between them with Micro Dissecting Knife-Needle (Fig. 2b, c).

19. Pool the testes and ovaries in the 1.5-ml tubes containing 500 μl RNAlater solution. Keep the tubes overnight at 4°C (see Note 5).

20. Add an equal volume pre-cooled PBS, and mix well.

21. Spin down at $5,000 \times g$ for 2 min and discard the supernatant. These gonads can be used for the isolation of RNA immediately or stored at −80°C until use.

3.3. Isolation of Gonads from Embryos Younger than 12.5 dpc

The major steps for dissecting gonads from embryos younger than 12.5 dpc are the same as Section 3.2. However, at these developmental stages, the urogenital ridges are morphologically indistinguishable between XX and XY embryos. If the sex of each gonad needs to be identified, an extra step is required for determining the sex of the embryos. Staining of Bar body in amnions is a quick method for identifying the XX embryo by the presence of condensed chromatin body (Bar body) and is particularly advantageous when the sex of gonad needs to be known before further procedures (10). Otherwise, a piece of embryonic body (e.g., tail) can be saved during dissection and used for genotyping later. PCR detection of Y-chromosome-specific sequences is accurate and convenient particularly when a large number of gonads must be collected. We describe here the modifications of the Section 3.2 for this purpose.

1. Follow Step 1 to 9 as described under Section 3.2

2. Place embryos in order, so that they can be identified.

3. Keep the tails of embryos in individual tubes marked with their identifications. Store them at −20°C until PCR genotyping.

4. Instead of pooling the gonads, isolate urogenital complexes or gonads from each embryo one by one and store them individually

in proper containers with either fixative or RNAlater solution (see Note 6).

3.4. Sexing the Embryos by PCR Amplification of the Sry Sequence

1. Add 50 µl DirectPCR Lysis Reagent containing freshly prepared 0.2 mg/ml Proteinase K to the 5 ml tube containing the tail of the embryo.

2. Incubate at 55°C for 5–6 h or overnight until no tissue clumps are seen.

3. Briefly spin down the tubes and add 50 µl H_2O to each tube.

4. Incubate the tubes at 95°C for 10 min to inactivate Proteinase K.

5. Use 1 µl of the lysate for total 25 µl PCR reaction containing a pair of *Sry* primers (Sry-BA: 5′-GTCAAGCGCCCCAT-GAATGC-3′;Sry-EA 5′-TAGTTTGGGTATTTCTCTCT-3′).

6. PCR conditions: 95°C for 5 min, 30 cycles of 1 min at 95°C, 1 min annealing at 55°C, 1 min extension at 72°C, and a final extension at 72°C for 10 min.

7. Run a 1.5% agarose gel to analyze the PCR products.

8. If a 203-bp fragment is amplified by PCR, the embryo is considered to be XY male; the absence of such a DNA fragment, the embryo is considered to be XX females.

3.5. Examples of Studies

1. Immunofluorescence detection of SOX9 expression in histological sections:

Testicular differentiation is morphologically distinct by the formation of testis cords in the XY gonad by 12.5 dpc. The involvement of various molecules in this process can be studied by immunofluorescence detection of proteins in histological sections of the gonads. The mesonephros is usually left attached to the gonad so that the orientation of gonad can be identified. We show here an example of double staining of SOX9 and mouse Vasa homolog (MVH). *Sox9* is a direct target of the primary testis-determining gene *Sry* (4) and represents Sertoli cell differentiation. MVH is a marker for the germ cells of both sexes (15).

Methods: Isolated urogenital complexes at 12.5 and 14.5 dpc are fixed in 2% paraformaldehyde in microtubule stabilizing buffer (3) for 1 h and embedded in paraffin. Slides from the mid part of gonads are de-paraffinized and subjected to antigen retrieval procedures as previously described (3). The slides are then incubated with the mouse IgG blocking reagent provided in the MOM kit (Vector Laboratories, Burlingame, CA) for 60 min, followed by an anti-SOX9 mouse monoclonal antibody (Abnova Laboratories, Taiwan) and a rabbit anti-MVH antibody (kind gift from Dr. T. Noce) both at 1:1,000 dilutions, in PBS containing the protein concentrate (provided in the MOM kit) at 4°C overnight. On the next day, all slides are washed with

PBS and incubated with a horse anti-mouse IgG antibody conjugated with biotin (provided in the MOM kit) and a goat anti-rabbit IgG antibody conjugated with Rhodamine (Pierce Endogen, Rochford, IL), both at 1:1,000 dilutions, in PBS containing the protein concentrate for 30 min. All slides are washed with PBS, followed by distilled water, and mounted in the Prolong Antifade mounting medium (Molecular Probe, Eugene, OR) containing 0.4 µg/ml 4′,6-diamidin-2′-phenylindole dihydrochloride (DAPI) (Roche Diagnostics, Mannheim, Germany). Fluorescent signals are examined under an epifluorescence microscope (Zeiss Axiophot, Germany). All images are captured with a digital camera (Retiga 1300, QImaging, Burnaby, BC) and processed with Northern Eclipse digital imaging software, version 6.0 (Empix Imaging, Mississauga, ON).

Results: SOX9 immunofluorescence staining can be clearly seen in the nuclei of numerous cells in the testes but not ovaries at either 12.5 or 14.5 dpc (Fig. 3). In the testis at 12.5 dpc, SOX9 staining is clearly seen in the nucleus. SOX9-positive cells are mainly localized within the testis cords but also occasionally seen in the interstitium, and less organized in the caudal pole. MVH staining is seen in the cytoplasm of germ cells. They are enclosed in SOX9-positive cells in the testis cords. By 14.5 dpc, both the volume of testis and number of testis cords increase considerably. SOX9-positive cells are predominantly localized within the testis cords, forming palisade-like alignments. In the ovary at 12.5 dpc, no SOX9 staining is seen. MVH-positive germ cells are scattered over the ovary. The number of MVH-positive cells is visibly increased in the ovary from 12.5 to 14.5 dpc.

2. MicroRNA profiling of 12.5 dpc fetal gonads

MicroRNA (miRNA)s are a class of short (~22 nt), naturally occurring, noncoding RNAs. They regulate protein expression by targeting the mRNAs of protein coding genes, resulting in either translation repression or transcriptional regulation (16). It has been shown that miRNAs are important for the proliferation of PGCs and spermatogonia. In addition, Dicer, the key enzyme of miRNAs biogenesis, is required for Sertoli cell functions and survival (17). Since miRNAs are important for organogenesis and cellular differentiation, it is reasonable to hypothesize that miRNAs also play some important roles in regulating fetal testis and ovary differentiation. Indeed, sexually dimorphic miRNA expression was observed during chicken fetal gonadal development (18). Accordingly, we used the isolated male and female gonads at 12.5 dpc for profiling the differentially expressed miRNAs.

Methods: Male and female gonads at 12.5 dpc from six CD1 mice are collected and pooled separately in 1.5-ml tubes containing RNAlater solution. Total RNA is isolated with RNAqueous®-

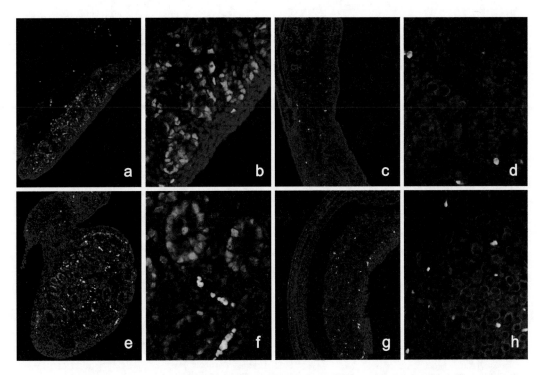

Fig. 3. Immunofluorescence detection of SOX9 (*green*) and MVH (*red*) with DAPI counterstaining (*blue*). The mesonephros is placed on the *left* and the gonad on the *right*. *Bar* indicates 200 μm in **a**, **c**, **e**, and **g**, or 40 μm in **b**, **d**, **f**, and **h**. SOX9 staining is predominantly seen in the nucleus of Sertoli cells and their precursors, whereas MVH staining is in the cytoplasm of germ cells. Yellow staining shows blood cells. (**a** and **b**) Male urogenital complex at 12.5 dpc. (**c** and **d**) Female urogenital complex at 12.5 dpc. (**e** and **f**) Male urogenital complex at 14.5 dpc. (**g** and **h**) Female urogenital complex at 14.5 dpc.

Micro Kit (Ambion) according to the manufacturer's instruction. RNA is quantified by using the NanoDrop ND-1000 Sepectrophotometer. Mouse mmu-miRNome MicroRNA Profiling Kit (System Biosciences), which detects the expression of 709 miRNAs in mouse, is used to profile the miRNAs in male or female gonads. 1 μg of total RNA is tagged with poly(A) to its 3′ end by poly A polymerase, and reverse-transcribed with oligo-dT adaptors by QuantiMir RT technology. Expression levels of the miRNAs are measured by quantitative PCR using SYBR green reagent and 7900HT Fast Real-Time PCR System (Applied Biosystems). All miRNAs can be measured with miRNAs specific forward primers and a universal reverse primer (SBI). Expression levels of the miRNAs can be normalized to U6 snRNA, RNU43 snoRNA, and U1 snRNA.

Results: Of the 709 miRNAs analyzed, 52 miRNAs were upregulated threefolds or higher in male gonads (Fig. 4a) and 69 miRNAs were upregulated at least threefolds in female gonads (Fig. 4b). The differential expression of miRNAs between male and female gonad at 12.5 dpc suggests that miRNAs may play important roles in sex differentiation. For the male gonads, the top ten miRNAs are miR-34b-5p,

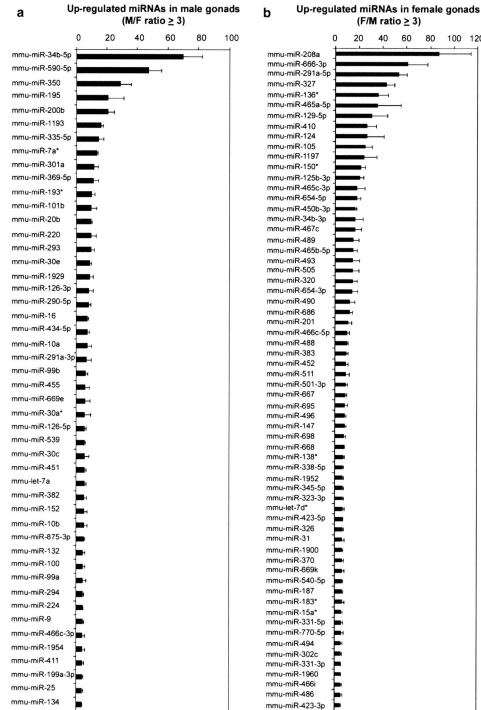

Fig. 4. Differential expression of miRNAs in mouse fetal gonads at 12.5 dpc. Total RNAs extracted from male and female gonads were used to measure the levels of miRNAs expression, and the ratio of male to female gonads was calculated. (**a**) miRNAs upregulated in male gonads as M/F ratios. (**b**) miRNAs upregulated in female gonads as F/M ratios. Data are derived from three independent experiments. Those showing ≥threefold changes are presented here.

miR-590-5p, miR-350, miR-195, miR-200b, miR-1193, miR-335-5p, miR-7a*, miR-301a, and miR-369-5p, while for the female gonads, the top 15 miRNAs are miR-208a, miR-666-3p, miR-291a-5p, miR-327, miR-136*, miR-465a-5p, miR-129-5p, miR-410, miR-124, miR-105, miR-1197, miR150*, miR125b-3p, miR-465c-5p, and miR-654-5p. Since the isolated gonads at 12.5-dpc stage contain various somatic cell types in addition to germ cells, the cellular origins and exact functions of these differentially expressed miRNAs in embryonic testis and ovary are uncertain. Nevertheless, other studies have suggested some possible roles of these miRNAs in regulation of gene expression and biological processes. Among the miRNAs upregulated in the embryonic testis, miR-34, miR-200, miR-195, and miR-7 are putative tumor suppressors (19, 20). miR-34b-5p has been demonstrated to mediate the expression of an alternatively polyadenylated variant of the mouse β-actin gene (21), which might be important for testis cord formation. The miR-200b, miR-20b, and miR-369-5p are involved in mesenchymal stem cell phenotype (22) and epithelial-mesenchymal transition (23). Further, miR-200b is associated with regulation of growth and function of stem cells, in which its upregulation is associated with differentiation while its inhibition is related to maintenance of stemness (24). miR-7a is highly expressed in testis-derived male germ stem cells (25) and could be derived from the male germ cells in the embryonic testis.

Among the miRNAs upregulated in the embryonic ovary, miR-208a represents one of the well-characterized miRNAs. It is encoded within an intron of α-cardiac muscle myosin heavy chain gene and is a regulator of cardiac hypertrophy and conduction (26, 27) via its regulation of a number of target genes, including GATA4 (26), correct dosage of which is needed for proper sex determination (28). miR-124 has been demonstrated to play critical role in neurogenesis (29, 30), particularly in its regulation of SOX9 expression in the developing nervous tissues (31). Hence, SOX9 could be a target for inhibition by miR-124 in ovarian cells, in which SOX9 expression is preferentially repressed, as compared to the testicular cells of the same stage. Interestingly, miR-105 has been shown to control human ovarian cell proliferation (32), its elevated expression in the fetal ovary suggests a potentially similar role in the mouse. Further detailed studies on the expression of miRNAs and identification of their corresponding targets at various stages of gonadal development in both male and female should shed critical insights on how these miRNAs regulate the sex-specific cell fate determination and organogenesis in this interesting developmental system.

4. Notes

1. Since embryos are very soft and fragile, gentle handling is required to avoid breaking the embryos, especially when you need to count the tail somites.

2. Do not add too much dissection medium to the dish during dissection. Otherwise, embryos will move around and be difficult for the dissection manipulations. It is much easier to position embryos on a sponge with a minimal less volume of medium.

3. In step 2, cut just below the liver. If cut too deep, the urogenital complex may be removed at this step.

4. We recommend 2% paraformaldehyde in microstabilizing buffer (3). Since this fixative is powerful, antigen retrieval is necessary prior to immunofluorescence procedures.

5. For RNA isolation from gonads of 12.5 dpc or older embryos, the gonads can also be snap-frozen in individual microcentrifuge tubes in liquid nitrogen at this step.

6. For RNA isolation from gonads of younger than 12.5 dpc, if the sex of the gonads is needed, it is convenient to place the gonads in 8-tube PCR strips containing RNAlater solution, which preserves the integrity of the RNAs for RNA purification procedures at a later time (per supplier of this reagent kit, Ambion Inc.) instead of being snap-frozen in liquid nitrogen.

Acknowledgments

This work was partially supported by an NIH grant to Y-FC Lau. Y-FC Lau is a Research Career Scientist in the Department of Veterans Affairs.

References

1. Wilhelm D, Palmer S, Koopman P (2007) Sex determination and gonadal development in mammals. *Physiol Rev* 87, 1–28.

2. Sekido R, Lovell-Badge R (2009) Sex determination and SRY: down to a wink and a nudge? *Trends Genet* 25, 19–29.

3. Taketo T, Lee CH, Zhang J, Li Y, Lee CY, Lau YF (2005) Expression of SRY proteins in both normal and sex-reversed XY fetal mouse gonads. *Dev Dyn* 233, 612–622.

4. Sekido R, Lovell-Badge R (2008) Sex determination involves synergistic action of SRY and SF1 on a specific Sox9 enhancer. *Nature* 453, 930–934.

5. Koopman P, Gubbay J, Vivian N, Goodfellow P, Lovell-Badge R (1991) Male development of chromosomally female mice transgenic for Sry. *Nature* 351, 117–121.

6. Tomizuka K, Horikoshi K, Kitada R, Sugawara Y, Iba Y, Kojima A, Yoshitome A, Yamawaki K, Amagai M, Inoue A, Oshima T, Kakitani M (2008) R-spondin1 plays an essential role in ovarian development through positively regulating Wnt-4 signaling. *Hum Mol Genet* 17, 1278–1291.

7. Kim Y, Kobayashi A, Sekido R, DiNapoli L, Brennan J, Chaboissier MC, Poulat F, Behringer

RR, Lovell-Badge R, Capel B (2006) Fgf9 and Wnt4 act as antagonistic signals to regulate mammalian sex determination. *PLoS Biol* 4, e187.

8. Uhlenhaut NH, Jakob S, Anlag K, Eisenberger T, Sekido R, Kress J, Treier AC, Klugmann C, Klasen C, Holter, NI, Riethmacher D, Schutz G, Cooney AJ, Lovell-Badge R, Treier M (2009) Somatic sex reprogramming of adult ovaries to testes by FOXL2 ablation. *Cell* 139, 1130–1142.

9. Taketo TTR, Adeyemo O, Koide SS (1984) Influence of adenosine 3′,5′-cyclic monophosphate analogs on testicular organizaion of fetal mouse gonads in virtro. *Biol Reprod* 30, 189–198.

10. Capel B, Batchvarov J (2008) Preparing Recombinant Gonad Organ Cultures. *Cold Spring Harb Protoc*

11. Li Y, Yue L, Taketo T, Lau YF (2003) Protein transduction as a strategy for evaluating important factors in mammalian sex determination and differentiation. *Cytogenet Genome Res* 101, 237–241.

12. Capel B, Albrecht KH, Washburn LL, Eicher EM (1999) Migration of mesonephric cells into the mammalian gonad depends on Sry. *Mech Dev* 84, 127–131.

13. Nef S, Schaad O, Stallings NR, Cederroth CR, Pitetti JL, Schaer G, Malki S, Dubois-Dauphin M, Boizet-Bonhoure B, Descombes P, Parker KL, Vassalli JD (2005) Gene expression during sex determination reveals a robust female genetic program at the onset of ovarian development. *Dev Biol* 287, 361–377.

14. Beverdam A, Koopman P (2006) Expression profiling of purified mouse gonadal somatic cells during the critical time window of sex determination reveals novel candidate genes for human sexual dysgenesis syndromes. *Hum Mol Genet* 15, 417–431.

15. Toyooka Y, Tsunekawa N, Takahashi Y, Matsui Y, Satoh M, Noce T (2000) Expression and intracellular localization of mouse Vasa-homologue protein during germ cell development. *Mech Dev* 93, 139–149.

16. Shah AA, Meese E, Blin N (2010) Profiling of regulatory microRNA transcriptomes in various biological processes: a review. *J Appl Genet* 51, 501–507.

17. Kim GJ, Georg I, Scherthan H, Merkenschlager M, Guillou F, Scherer G, Barrionuevo F (2010) Dicer is required for Sertoli cell function and survival. *Int J Dev Biol* 54, 867–875.

18. Bannister SC, Tizard ML, Doran TJ, Sinclair AH, Smith CA (2009) Sexually dimorphic microRNA expression during chicken embryonic gonadal development. *Biol Reprod* 81, 165–176.

19. O'Day E, Lal A (2010) MicroRNAs and their target gene networks in breast cancer. *Breast Cancer Res* 12, 201.

20. Liu L, Chen L, Xu Y, Li R, Du X (2010) microRNA-195 promotes apoptosis and suppresses tumorigenicity of human colorectal cancer cells. *Biochem Biophys Res Commun* 400, 236–240.

21. Ghosh T, Soni K, Scaria V, Halimani M, Bhattacharjee C, Pillai B (2008) MicroRNA-mediated up-regulation of an alternatively polyadenylated variant of the mouse cytoplasmic {beta}-actin gene. *Nucleic Acids Res* 36, 6318–6332.

22. Giraud-Triboult K, Rochon-Beaucourt C, Nissan X, Champon B, Aubert S, Pietu G (2011) Combined mRNA and microRNA profiling reveals that miR-148a and miR-20b control human mesenchymal stem cell phenotype via EPAS1. *Physiol Genomics* 43,77–86.

23. Castilla MA, Moreno-Bueno G, Romero-Perez L, De Vijver KV, Biscuola M, Lopez-Garcia MA, Prat J, Matias-Guiu X, Cano A, Oliva E, Palacios J (2011) Micro-RNA signature of the epithelial-mesenchymal transition in endometrial carcinosarcoma. *J Pathol* 223, 72–80.

24. Iliopoulos D, Lindahl-Allen M, Polytarchou C, Hirsch HA, Tsichlis PN, Struhl K (2010) Loss of miR-200 inhibition of Suz12 leads to polycomb-mediated repression required for the formation and maintenance of cancer stem cells. *Mol Cell* 39, 761–772.

25. Jung YH, Gupta MK, Shin JY, Uhm SJ, Lee HT (2010) MicroRNA signature in testes-derived male germ-line stem cells. *Mol Hum Reprod* 16, 804–810.

26. Callis TE, Pandya K, Seok HY, Tang RH, Tatsuguchi M, Huang ZP, Chen JF, Deng Z, Gunn B, Shumate J, Willis MS, Selzman CH, Wang DZ (2009) MicroRNA-208a is a regulator of cardiac hypertrophy and conduction in mice. *J Clin Invest* 119, 2772–2786.

27. van Rooij E, Quiat, Johnson BA, Sutherland LB, Qi X, Richardson JA, Kelm RJ Jr, Olson EN (2009) A family of microRNAs encoded by myosin genes governs myosin expression and muscle performance. *Dev Cell* 17, 662–673.

28. Bouma GJ, Washburn LL, Albrecht KH, Eicher EM (2007) Correct dosage of Fog2 and Gata4 transcription factors is critical for fetal testis development in mice. *Proc Natl Acad Sci USA* 104, 14994–14999.

29. Cheng LC, Pastrana E, Tavazoie M, Doetsch F (2009) miR-124 regulates adult neurogenesis in the subventricular zone stem cell niche. *Nat Neurosci* 12, 399–408.

30. Gao FB (2010) Context-dependent functions of specific microRNAs in neuronal development. *Neural Dev* 5, 25.

31. Farrell BC, Power EM, Dermott KW (2011) Developmentally regulated expression of Sox9 and microRNAs 124,128 and 23 in neuroepi-thelial stem cells in the developing spinal cord. *Int J Dev Neurosci* 29, 31–36.

32. Sirotkin AV, Laukova M, Ovcharenko D, Brenaut P, Mlyncek M (2009) Identification of micro RNAs controlling human ovarian cell proliferation and apoptosis. *J Cell Physiol* 223, 49–56.

Chapter 2

Neonatal Testicular Gonocytes Isolation and Processing for Immunocytochemical Analysis

Gurpreet Manku, Monty Mazer, and Martine Culty

Abstract

In recent years, an increasing interest has emerged at understanding how spermatogonial stem cells (SSCs) arise from their precursor cells, the gonocytes. The identification of factors acting directly on gonocytes rather than on the surrounding somatic cells and the study of genes and signaling pathways intrinsic to gonocytes require their isolation from Sertoli and peritubular cells. The present article describes a simple method developed to isolate rat neonatal gonocytes, allowing for the study of their proliferation and differentiation to SSCs. We also present immunocytochemical methods that can be used to study protein expression changes and proliferation in gonocytes.

Key words: Testicular gonocytes, Primary germ cell isolation, BSA gradient, Nonadherent cell immunocytochemistry

1. Introduction

Spermatogenesis relies on the formation of an appropriate niche of spermatogonial stem cells (SSCs) between gestation and infancy that will provide constant starting material for the generation of spermatozoa throughout the life of an individual (1). Recent findings that SSCs can be reprogrammed to behave as pluripotent embryonic stem cells (2), together with an unexplained increase in testicular germ cell tumors (3), have led to a heightened interest in understanding how SSCs form from their precursor cells, the gonocytes, also referred to as pre- or prospermatogonia. Rodent models have been routinely used to study germ cell development because they present many similarities with their human counterpart while offering the advantage of recapitulating within a few weeks a process that normally takes years to occur. Our goal was to establish a

Wai-Yee Chan and Le Ann Blomberg (eds.), *Germline Development: Methods and Protocols*, Methods in Molecular Biology, vol. 825, DOI 10.1007/978-1-61779-436-0_2, © Springer Science+Business Media, LLC 2012

simple method to isolate and keep in culture rat neonatal germ cells that could be further used to study the regulatory mechanisms involved in their proliferation and differentiation to SSCs. Although newer methods have emerged in the past years that is fluorescence-activated cell sorting (FACs) and magnetic-activated cell sorting (MACs) to generate enriched SSC preparations (4), these methods have not been successful in isolating sufficient numbers of gonocytes due to their heterogeneity of phenotypes. Gonocyte development takes place within a narrow time-frame in a non-synchronized manner. As a result, one can find germ cells presenting different functional and expression profiles during perinatal development (5). The isolation method presented here corresponds to a modified and miniaturized version of the original STAPUT gradient method developed by Romrell et al. (6). We also describe methods used to visualize protein expression or cell proliferation by immunocytochemistry (ICC) in gonocytes.

This method is designed to obtain an enriched population of gonocytes from testes where germ cells represent less than 1% of the total cells. The cells are routinely isolated from the pooled testes of 30 postnatal day (PND) 3 Sprague Dawley rat pups (7). The principle of the method is to first isolate intact seminiferous cords from the interstitium by disrupting the interstitium with collagenase and hyaluronidase treatment, followed by further dissociation of the tubules with trypsin. A large portion of the somatic cells is then removed during an overnight culture of the cell suspension, during which the somatic cells adhere to the culture plate while germ cells remain non-adherent. Further gonocyte enrichment is achieved by separating the large germ cells from the much smaller somatic cells by sedimentation velocity at unit gravity using a bovine serum albumin (BSA) gradient. Depending on the level of gonocyte purity needed and the minimum number of cells required by the experiments, an additional level of enrichment can be achieved by collecting cells using a micromanipulator microscope. Because gonocytes are non-adherent cells available in limited amounts, ICC represents a method of choice to study changes in protein expression and/or activation as well as proliferation in these cells. Thus, we describe the process used to perform ICC on gonocytes.

2. Materials

2.1. Tissue Dissociation and Cell Isolation

1. All reagents have to be sterile or sterilized before use.
2. RPMI 1640 (RPMI) (Invitrogen) containing 100 U/ml penicillin and 100 mg/ml streptomycin (CellGro) supplemented with 0, 5, or 10% heat-inactivated Fetal Bovine Serum (FBS) (Invitrogen) (see Note 1).

3. Collagenase: 2 mg/ml RPMI (containing antibiotics), sterilize using 0.2-μm filters and freeze aliquots at –20°C.

4. Hyaluronidase: Type III testicular Hyaluronidase: 1 mg/ml RPMI (containing antibiotics), sterilize using 0.2-μm filters and freeze aliquots at –20°C.

5. DNAse: 1 mg/ml RPMI (containing antibiotics), sterilize using 0.2-μm filters and freeze aliquots at –20°C.

6. Collagenase–Hyaluronidase–DNAse solution: 4.6 ml RPMI (containing antibiotics but no serum), 2.5 ml collagenase, 800 μl DNAse, and 100 μl hyaluronidase.

7. Trypsin–DNAse I solution: 3 ml commercial sterile solution of 0.25% Trypsin+1 mM ethylenediamine tetraacetic acid (EDTA) and 0.3 ml DNAse I stock (see Subheading 2.5) added.

2.2. BSA Gradient

1. 2% BSA solution: 0.6 g BSA fraction V (Roche) in 30 ml serum-free RPMI with antibiotics (filter solution using 0.2-μm filter).

2. 4% BSA solution: 1.2 g BSA fraction V (Roche) in 30 ml serum-free RPMI with antibiotics (filter solution using 0.2-μm filter).

3. Metricide (CSR): add 2.65 g Metricide powder to 500 ml solution.

4. The soultions can be kept overnight at 4°C.

2.3. Cytospin and Immunocytochemistry

1. Paraformaldehyde solution:

(a) Stock 8% Paraformaldehyde solution:

- Dissolve 16 g of paraformaldehyde in 80 ml of double-distilled (dd) water by stirring at 70°C in the fume hood.

- Add few drops on 1N NaOH to depolymerize the para-formaldehyde (otherwise, the powder never dissolves).

- Adjust the pH to 7.0 with 1N HCl by using pH paper. With a Pasteur pipette, take out a drop of the solution and put it on the pH paper to verify the pH. Do not soak the pH paper in the solution to avoid coloring the solution.

- Cool down to room temperature. Bring up to 100 ml with dd water.

- Filter through Whatman paper.

- Mix with an equal volume of 2× PBS (e.g., 100 ml).

- Divide into convenient aliquots and store frozen at –20°C.

(b) Paraformaldehyde 4% working solution:

- Thaw the volume of paraformaldehyde stock solution you will need.

- Add 1 volume of 1× PBS to make it a 4% paraformal-dehyde working solution.

- Discard any leftover working solution in the hazard-ous chemical waste.

2. Other reagents:

   (a) 10 μM Protease inhibitor cocktail: 1 tablet (Roche) in 10 ml water.

   (b) 10 μM Phosphatase inhibitor: 1 tablet (Roche) in 10 ml water, 4 μM okadaic acid: use solution as is (Calbiochem) (see Note 2).

   (c) DAKO antigen retrieval solution: 1 part 10× Dako Target Retrieval solution in 9 parts water.

   (d) Blocking solution: 1× PBS, 1% BSA, 10% goat serum.

   (e) Peroxide solution: 7.5 ml 100% H_2O_2, 225 ml methanol, 17.5 ml H_2O.

   (f) Primary antibody solution: 1× PBS, 1% BSA, 2% goat serum, 0.02% Triton X-100.

   (g) Primary antibodies: most antibodies are used at 1:100 dilutions. Dilution can be adjusted for antibodies with weaker or stronger affinity to their antigens.

   (h) Fluorescent secondary antibody solution: 1× PBS, 1% BSA, 0.02% Triton X-100.

   (i) Fluorescent secondary antibodies: Goat anti-Rabbit and Goat anti-Mouse Antibodies coupled to Alexa 488, 546 (usually, a dilution of 1:300 is used).

   (j) Colorimetric secondary antibody solution: 1× PBS, 1% BSA.

   (k) Colorimetric secondary antibodies: Biotin Goat Anti-Rabbit and Biotin Goat Anti-Mouse (BD Pharmingen).

   (l) Nuclear Staining: Hoechst 33342 (Molecular Probes) stock solution is 10 mg/ml (16.2 mM) in DMSO. Take 2 μl of this stock and add it to 10 ml 1× PBS; DAPI (Invitrogen) 14.3 mM Stock, 600 nM working concentration, 5 μL stock in 500 μL PBS (×100); 5 μL Solution in 1.2 mL PBS (×240).

   (m) Streptavidin/Peroxidase (Invitrogen): ready-to-use solution.

   (n) Single solution chromogen of 3-amino-9- ethylcarbazole (AEC; Invitrogen): ready-to-use solution.

   (o) Hematoxylin: ready-to-use solution.

   (p) PermaFlour (ThermoScientific): ready-to-use solution.

3. Methods

3.1. Tissue Collection

1. Testes collection: 30–40 male neonatal Sprague Dawley rats are purchased at PND2 (day of birth is day zero) from Charles Rivers Laboratories, who sends them in cages of 10 pups with a foster mother (see Note 3).

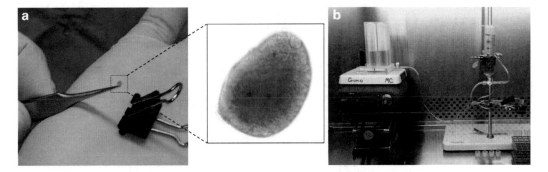

Fig. 1. Illustrations of a neonatal rat testis and BSA gradient device. (**a**) the size and appearance of a rat PND3 testis are shown. Insert: whole testis observed with the 40× objective using an inverted microscope. (**b**) BSA gradient settings. The height of the gradient former and the syringe where the gradient is formed are adjusted to obtain a slow flow-rate of liquid. Fraction collection can be performed with a fraction collector or manually into sterile tubes.

2. Because neonatal rats are resistant to CO_2 asphyxia, PND3 pups are euthanized by inducing hypothermia and decapitation according to the Guide for the Care and Use of Experimental Animals from the Canadian Council on Animal Care and McGill University.

3. Use sterile conditions throughout the procedure. Prepare a sterile area and use instruments sterilized by autoclaving.

4. Prepare 30 ml RPMI (containing antibiotics) in a 50-ml tube, keep in ice for use in collecting and decapsulating testis in steps 6 and 7.

5. Thaw the enzymes (collagenase, hyaluronidase, and DNAse), 10% FBS, and Trypsin–DNAse solution for the tissue dissociation.

6. To collect testes, spray 70% ethanol on the body of the rat pup and perform an abdominal incision. Neonatal testes are positioned in the lower abdomen on either sides of the bladder. They are easily identified by their ovoid shape (see Fig. 1a). A binocular dissecting microscope can be used to facilitate the procedure if needed. (Testes are kept in a 50-ml tube of RPMI on ice for the duration of the dissection).

7. Once collected, testes are cleaned from the epididymis and other surrounding tissues. Testes are then decapsulated, pooled in a 50-ml tube with RPMI with antibiotics, and processed for tissue dissociation.

8. Continue the experiment in a culture hood under sterile conditions.

3.2. Tissue Dissociation

1. Collagenase–hyaluronidase–DNAse treatment: Replace RPMI above the decapsulated testes with Collagenase–Hyaluronidase–DNAse solution (see Subheading 2.1, item 6).

2. Incubate testes for 30 min at 37°C in a shaking water bath (position tube in a way that allows efficient shaking of the tissues and maintenance of sterility).

3. Remove as much supernatant as possible without disturbing the tissue pellet. The supernatant contains interstitial cells, including Leydig cells, endothelial, and blood cells. (This suspension can be further used to isolate Leydig cells. In that case, add RPMI containing 10% FBS, centrifuge, and apply the resuspended cell pellet onto a Percoll gradient). The tissue pellet that remains contains mainly seminiferous tubules (called cords in fetuses and neonates). Proceed to gonocyte isolation.

4. Trypsin–DNAse I treatments: Add Trypsin–DNAse I solution to the tissue pellet. Incubate 10–15 min at 37°C in the shaking water bath. Gently flush 4–5 times with pipette to perform mechanical dissociation (see Note 4).

5. Collect the supernatant into a tube containing 20 ml RPMI (containing antibiotics and 10% FBS as trypsin action can be inhibited using serum). Filter cell mixture through a 40-μm sterile nylon mesh to remove fragments. Keep on ice.

6. If a large tissue pellet remains after the first trypsin digestion, add 3 ml Trypsin–DNAse I solution and incubate further for 5–10 min at 37°C in the shaking water bath. Flush 2–3 times with pipette.

7. Collect the supernatant into a tube containing 20 ml RPMI (containing antibiotics and 10% FBS). Filter cell mixture through 40-μm sterile nylon mesh to remove fragments. Keep on ice.

8. Pool the two filtered cell suspensions and centrifuge for 5 min at $800 \times g$ to pellet down the cells.

9. Remove the supernatant and add 10 ml RPMI (containing antibiotics and 5% FBS) to the resulting cell pellet.

10. Count the cells using a hemocytometer.

11. Count the cells in 4×16 squares. There are 2 types of cells: Large cells (gonocytes) and small cells (somatic cells: ~70% Sertoli cells; 30% myoid cells).

12. Calculation: Number of cells in 16 squares $\times 10^4$ = number of cells/ml. Multiply by total volume (10 ml).

13. In general, one will get ~$100–200 \times 10^6$ somatic cells and $1–2 \times 10^6$ gonocytes from 30 pups.

3.3. Overnight Adhesion

1. Aliquot cell suspension in five large culture plates (150 mm) using 30 ml RPMI (containing antibiotics and 5% FBS) per plate.

2. Incubate plates overnight in a 3.5% CO_2 incubator at 37°C.

3. Use 6–8 plates for an experiment with 40 pups (see Note 5).

4. The somatic cells adhere to the plates, while gonocytes stay non-adherent. This step allows the removal of approximately half of the somatic cells from the initial cell suspension.

3.4. Cell Separation on a BSA Gradient

1. BSA (2–4%) gradient preparation: All equipment, gradient former and tubing must be sterilized using a Metricide disinfectant solution (CSR) overnight. Rinse all equipment with sterile water and RPMI before using. This is a scaled-down version of the traditional glass STAPUT gradient device. The gradient former is from BioRad (normally sold for the preparation of gradient polyacrylamide gels).

2. Connect the output stopcock of the gradient former to a flexible tube ending on a 3-way-stopcock attached to a 60 ml sterile syringe (in which the gradient will be formed). Another tube connects the third port of the syringe to a fraction collector (see Fig. 1b).

3. In each of the two chambers of the gradient former, add 30 ml of 2% or 4% BSA solution, placing the 4% BSA in the back chamber. Add a magnetic bar in each chamber and place the gradient former on a magnetic stir plate to activate the magnets (which prevents bubble formation in the gradient).

4. Open the connection between the back and front chambers and the connection linking the gradient former to the syringe. Also, open the 3-way-stopcock to let the gradient slowly form in the syringe. Use a slow flow rate to avoid disruption of the gradient.

5. Cell separation: While the gradient is forming, collect all supernatants containing gonocytes from the large culture plates. Gently flush the cell layers with RPMI to recover settled gonocytes. Centrifuge cells for 5 min at $800 \times g$.

6. Remove the supernatant and resuspend the cell pellet in 3 ml RPMI (containing antibiotics but no serum).

7. Count the number of gonocytes and small cells using the hemocytometer. Cell viability can be checked by trypan blue exclusion if needed.

8. Once the cells have been collected and the gradient has been formed, load the cell suspension slowly on top of the BSA gradient.

9. Let the cells settle down for 2.25 h. The cells settle in the gradient by gravity.

10. Collect 30–40 fractions of 2 ml each at a flow rate of 1.5 ml/2 min.

11. Count the gonocytes and small cells using the hemocytometer for all fractions (see Note 6).

12. Once the system is well standardized, one can collect fractions 1–10 in a single tube. Later fractions have to be assessed individually to determine the limits of the gonocyte peak.

13. Pool the fractions containing the most gonocytes and the least number of small cells (usually between fractions 1–18 to get 75–85% gonocyte purity).

14. Estimate gonocyte enrichment/purity of the final cell suspension by counting gonocytes and small cells using the hemocytometer. From 30 pups, one can usually obtain $0.15–0.3 \times 10^6$ enriched gonocytes.

15. In general, gonocytes are cultured at 10,000 cells/well in 150–500 μl RPMI (containing antibiotics but no serum), in function of the assay to perform and of the length of incubation period.

16. If a higher purity is required, perform further cell selection by micromanipulation. This allows to reach an ~100% purity, but it is very time-consuming and yields to a smaller number of gonocytes.

3.5. Micromanipulation

1. Micromanipulation is individual cell sorting performed on the gonocyte-rich pool from the BSA gradient. A micromanipulator system (TransferMan NK micromanipulator from Eppendorf Scientific; Brinkmann Instruments, Inc., Westbury, NY) is coupled to an Olympus inverted microscope, equipped with a cell harvesting system coupled to a 15-μm glass transfer micropipette.

2. Divide the gonocyte suspension between several 60-mm plates. Take only one plate at a time out of the incubator for cell collection because one does not want to leave cells at room temperature for a longer period of time than necessary.

3. Gonocytes are visualized on the monitor and selectively collected by aspiration in the transfer micropipette. Small batches of gonocytes are released in a 35-mm plate containing RPM (containing antibiotics).

4. Centrifuge the final pool of pure gonocytes (3,000–6,000 cells per sorting/~3 h) and freeze at −80°C for further use (usually RNA or protein extraction).

3.6. Cell Fixation and Cytospin Centrifugation

Because gonocytes are non-adherent cells, they need to be collected on microscopic slides for ICC analysis. ICC can be used to study the expression and/or activation of specific proteins using antibodies raised against specific proteins or phospho-proteins. This method can also be used to measure gonocyte proliferation in response to specific factors or inhibitors, either by determining Bromodeo-xyuridine (BrdU) incorporation or by following Proliferating Cell Nuclear Antigen (PCNA) expression in the cells with specific antibodies.

1. Cell fixation: At the end of the incubation, add 8% fresh paraform-aldehyde directly onto the cell suspension at a 1:1 ratio with the

medium to get a final concentration of 4% paraformaldehyde. This step helps preserve cell morphology during cytospin centrifugation (see Note 7).

2. Shake the tubes and leave for 7 min at room temperature.

3. Centrifuge the fixed cell suspension at $2,600 \times g$ for 5 min to decrease the volume that will be applied into the cytospin funnels.

4. Remove the supernatant and resuspend the cell pellet in 150 μl 1× PBS. If more than 1 slide will be prepared from a sample, add enough PBS to allow pipetting 100–150 μl per slide. One should aim for around 15,000 gonocytes per 100–150 μl.

5. Cytospin centrifugation (Shandon cytospin 3; Shandon Scientific Ld, Pittsburgh, PA): Set up slides in the cytospin machine and add the gonocyte suspension carefully into the bottom of the funnels.

6. Centrifuge for 10 min at $11 \times g$.

7. Dry slides in chemical hood for 1 h.

8. Fix cells in Acetone–Methanol (60:40) for 7 min.

9. Dry in chemical hood for 30 min or overnight at 4°C (cover with foil to protect from dust).

10. Wrap slides in foil and store at –20°C.

3.7. Immunocytochemical Analysis up to First Antibody Reaction

1. Take cytospin slides out of the freezer and let them defrost.

2. Draw a circle on the bottom of the slide around the sample with a diamond pen.

3. Heat DAKO antigen retrieval solution to 95°C in a large beaker on the hot plate.

4. Place the slides in the metal holder and keep on the hot plate for 15 min. Then, take the holder out and let it sit on the bench top for 10 min.

5. Remove slides from metal holder and put them in a staining Coplin jar.

6. Wash the slides for 5 min in 1× PBS + 0.1% Triton X-100 at room temperature.

7. Wash the slides for 5 min in 1× PBS at room temperature.

8. Soak the slides for 10 min in the peroxide solution.

9. Wash the slides for 5 min in 1× PBS + 0.1% Triton X-100 at room temperature.

10. Wash the slides for 5 min in 1× PBS at room temperature.

11. Use Kimwipes to wipe away the excess liquid off the slides.

12. Using a DAKO pen, make a circle, larger than the circle made by the diamond pen, to insure that the solutions will stay on top of the cells.

13. Place the slides in a humidified chamber. The chamber can be any plastic container with a sealable cover containing a large and flat sponge. Make sure to dampen the sponge with water to keep the box humidified.

14. Apply the blocking solution for 1 h at room temperature (approximately 100 μl is needed to cover the surface of a sample on a cytospin slide).

15. Remove the liquid from the slides.

16. Wash the slides for 5 min in 1× PBS at room temperature.

17. Wipe carefully around the samples on the slides using Kimwipes.

18. The dilution wanted for most primary antibodies is 1:100.

19. Add approximately 100 μl to each slide.

20. Incubate the slides at 4°C overnight in the humidified chamber.

21. The next day, wash the slides for 5 min in 1× PBS at room temperature.

22. Wipe carefully around the samples on the slides.

23. Up until now, everything is the same whether fluorescent staining or colorimetric staining is being done. It is at this step that both methods become different.

3.8. Fluorescent Immunocyto-chemical Analysis

1. At this point, everything has to be protected from the light. Wrap the plastic container in aluminum foil.

2. Apply the secondary (fluorochrome-conjugated) antibody for 1 h at room temperature (approximately 100 μl is needed for each cytospin).

3. The main dilution used for the secondary antibody is 1:300.

4. Add 100 μl of the secondary antibody to each slide.

5. After 1 h, wash the slides for 5 min at room temperature two times with 1× PBS.

6. Wipe carefully around the slide and apply 100 μl of the Hoechst solution or DAPI solution for ONLY 5 MIN (if left any longer, the nuclear staining will be too intense).

7. Wash the slides for 5 min at room temperature two times with 1× PBS.

8. Mount the slides with 1 drop of room temperature PermaFluor Mountant (Thermo Scientific). Make sure there are no air bubbles on the slide and then carefully place the coverslip on the slide. Let the solution dry at room temperature for 30 min.

9. Store the slides at −20°C protected from the light until you are ready to view them under the microscope.

10. It is highly recommended to look at the slides as soon as possible because the fluorescent signal will fade and will no longer be visible after 2–3 days.

3.9. Colorimetric Immunocytochemical Analysis

1. Apply the biotinylated secondary antibody for 1 h at room temperature (approximately 100 μl is needed for each cytospin).

2. The dilution wanted for the secondary antibody is usually 1:100.

3. Add 100 μl of the secondary antibody to each slide.

4. After 1 h, wash the slides for 5 min at room temperature two times with 1× PBS.

5. Place 2–3 drops (enough to cover the cells) of Streptavidin-Peroxidase (Invitrogen) on each slide.

6. Let sit at room temperature in the humidified chamber for 10 min.

7. Wash the slides with 1× PBS for 5 min, two times.

8. Using a plastic drop pipette, place 2–3 drops (enough to cover the cells) of AEC Single Solution Chromogen (Invitrogen) on each slide (Do NOT equilibrate the solution to room temperature before use).

9. Let slides sit at room temperature in the humidified chamber for 10–15 min.

10. Wash the slides once with distilled water for 5 min.

11. Using a plastic pipette, place 2–3 drops of Hematoxylin (Invitrogen) (enough to cover the cells) on the slides.

12. Incubate this at room temperature for 3 min.

13. Wash the slides with tap water for 5 min, two times.

14. Let the slides air dry slightly; remove excess liquid using Kimwipes.

15. Put a few drops of Crystal Mount onto the slides and make sure that there is a thin layer covering the whole sample. Also, make sure that there are no bubbles and that the layer of Crystal Mount is relatively uniform and not too thick.

16. Incubate at 80°C for 10 min or until fully dried.

17. Place a few drops of room temperature PermaFluor (Thermo-Scientific) onto the slides and carefully cover the cells with a coverslip, making sure that there are no air bubbles and that the coverslip covers the whole sample. Let dry for 30 min.

18. The slides can now be viewed under the microscope and can be stored in a slide box at room temperature. See Fig. 2 for representative ICC results of isolated gonocytes using fluorescent and colorimetric reactions.

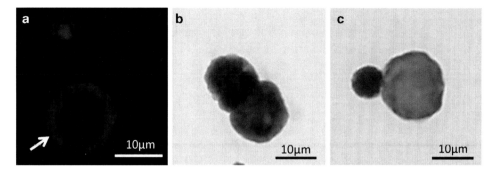

Fig. 2. Examples of immunocytochemical reactions. Fluorescent (**a**) and colorimetric (**b**) immunocytochemical reactions were performed on PND3 gonocytes collected by cytospin centrifugation. (**a**) estrogen receptor β expression in gonocytes incubated for 60 min with serum-free medium. Note the size difference between gonocyte (*arrow*) and somatic cell. (**b, c**) examples of PCNA expression in gonocytes cultured for 24 h. The picture in B shows proliferating PCNA-positive gonocytes. The gonocyte shown in C is quiescent and PCNA-negative, adjacent to a proliferative PCNA-positive somatic cell.

4. Notes

1. For cell isolation carried out during the summer, the addition of 0.25 mg/ml Amphotericin B (CellGro) may be useful to prevent yeast contamination.

2. If protein phosphorylation is studied, one can add a cocktail of phosphatase inhibitors (including okadaic acid, which is cell permeable) and protease inhibitors (leupeptin, PMSF etc.) right before adding paraformaldehyde onto the cells and within the formaldehyde solution to prevent dephosphorylation.

3. Although the yields can be improved by using a few more pups, we have found that keeping the dissection time to a minimum is critical for good cell preparations. Thus, taking too long to dissect more animals might hamper the benefit of using more rats.

4. A light mechanical dissociation simultaneously at the end of trypsinization will help. However, do not flush the tissue too much with the pipette or you will get viscous tissue aggregates and a poor yield of cell recovery after trypsinization.

5. Cells should not be too crowded during the o/n incubation as the Sertoli/myoid cells will entrap gonocytes to the bottom of the plates if the cells are crowded.

6. Gonocytes are easily distinguished from somatic cells. They are 2–3 times larger than somatic cells and are circular and shinny.

7. To collect gonocytes after a long incubation period (e.g., overnight proliferation assay) or if one of the contents (e.g., BSA) of the medium increases the risk of cross-linking the formaldehyde and aggregate cells, one can pellet the cells first by centrifugation

at $800 \times g$ for 5 min, discard the supernatant, wash once with 1× PBS if needed, and add 200 µl of 4% paraformaldehyde solution directly onto the cell pellet. Proceed then as described above in Subheading 3.6, step 4.

Acknowledgments

This work was supported in part by funds from The Research Institute of the McGill University Health Center and the Center for the Study of Reproduction, McGill University. The Research Institute of MUHC is supported in part by a Center grant from Le Fonds de la Recherche en Santé du Quebec.

References

1. de Rooij DG, Russel LD (2000) All you wanted to know about spermatogonia but were afraid to ask. Journal of Andrology 21:776–798

2. Guan K, Nayernia K, Maier LS et al. (2006) Pluripotency of spermatogonial stem cells from adult mouse testis. Nature 440:1199–1203

3. Reuter VE (2005) Origins and molecular biology of testicular germ cell tumors. Modern Pathol 18:S51–S60

4. Hofmann MC, Braydich-Stolle L, Dym M (2005) Isolation of male germ-line stem cells; influence of GDNF. Dev Biol 279:114–124

5. Culty M (2009) Gonocytes, the forgotten cells of the germ cell lineage. Birth Defects Research Part C 87:1–26

6. Romrell LJ, Bellvé AR, Fawcett DW (1976) Separation of mouse spermatogenic cells by sedimentation velocity. A morphological characterization. Dev Biol 49:119–131

7. Thuillier R, Mazer M, Manku G, Boisvert A, Wang Y, Culty M (2010) Interdependence of PDGF and estrogen signaling pathways in inducing neonatal rat testicular gonocytes proliferation. Biology of Reproduction 82:825–836

Chapter 3

Isolation of Undifferentiated and Early Differentiating Type A Spermatogonia from Pou5f1-GFP Reporter Mice

Thomas Garcia and Marie-Claude Hofmann

Abstract

Limited understanding of the mechanisms underlying self-renewal and differentiation of spermatogonial stem cells hampers our ability to develop new therapeutic and contraceptive approaches. Mouse models of spermatogonial stem cell development are key to developing new insights into the biology of both the normal and diseased testis. Advances in transgenic reporter mice have enabled the isolation, molecular characterization, and functional analysis of mouse Type A spermatogonia subpopulations from the normal testis, including populations enriched for spermatogonial stem cells. Application of these reporters both to the normal testis and to gene-deficient and over-expression models will promote a better understanding of the earliest steps of spermatogenesis, and the role of spermatogonial stem cells in germ cell tumor.

Key words: Reporter mice, Undifferentiated spermatogonia, Pou5f1, Green fluorescent protein, Fluorescence-activated cell sorting

1. Introduction

The use of GFP (green fluorescent protein)-reporter mice for the in vivo and ex vivo identification of specific cells from a heterogeneous population is unsurpassed in terms of ease and accuracy. GFP reporter mice that have been successfully developed for the identification of postnatal germ cells include Dazl-GFP (1), Neurog3-GFP (2), Stra8-GFP (3), and Pou5f1-GFP (4). Of these, the Pou5f1-GFP reporter mouse is unique in that it allows visualization of Pou5f1 positive (+) cells (5, 6), which in adult males marks early Type A spermatogonia (7, 8). Pou5f1+ cells may be further divided based on their expression of the cell surface receptor, Kit (also referred to as c-Kit) (4). As Kit is a marker for differentiating spermatogonia (9), Pou5f1+/Kit– cells comprise undifferentiated

Wai-Yee Chan and Le Ann Blomberg (eds.), *Germline Development: Methods and Protocols*, Methods in Molecular Biology, vol. 825, DOI 10.1007/978-1-61779-436-0_3, © Springer Science+Business Media, LLC 2012

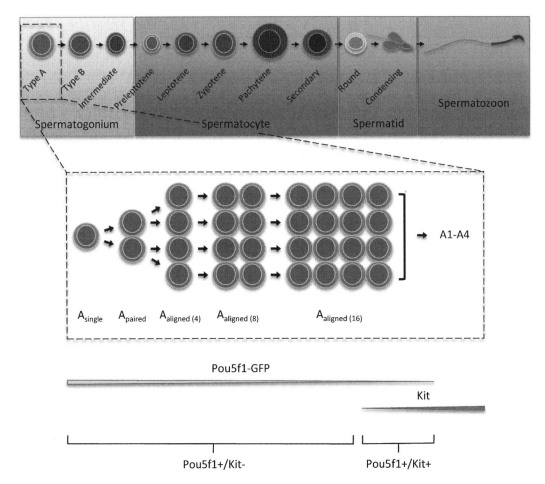

Fig. 1. Schematic diagram of spermatogenesis in the prepubertal and adult mouse testis showing the expression time-course of Pou5f1 and Kit. Spermatogenesis begins with mitotic proliferation and differentiation of diploid, undifferentiated spermatogonia. Differentiating spermatogonia unidirectionally continue differentiation into spermatocytes, which undergo two meiotic divisions to form haploid spermatids. Haploid spermatids undergo dramatic structural modifications leading up to the formation of spermatozoa. The time from the initiation of differentiation to the formation of spermatozoa is a period of several weeks. Day 7 testis only contain spermatogenic cells up to Type B spermatogonia; Day 10 testis only contain spermatogenic cells up to leptotene spermatocytes (28). Diagrams do not depict the relative proportion of each cell population.

spermatogonia, whereas Pou5f1+/Kit+ cells are constituents of the earliest steps of differentiation (Fig. 1). Therefore, Pou5f1+ cells as a whole are a mixture of undifferentiated and early differentiating Type A spermatogonia. This notion is supported by transplantation studies that have shown Pou5f1+/Kit+ cells are inefficient at colonizing a germ-cell depleted testis relative to Pou5f1+/Kit– transplanted cells (4, 10, 11).

Our intention here is not to review Pou5f1 (also referred to as Oct4) as an embryonic stem (ES) cell and pluripotency marker, nor go into great detail describing the constructs used to create the transgenic reporter strains based on the Pou5f1 gene promoter.

Since several publications and reviews effectively cover these and many other aspects (10, 12, 13), rather we present here several features relevant to the use of the Pou5f1-GFP reporter mouse for postnatal Type A spermatogonia identification and isolation (Fig. 2). In particular, we describe fluorescence-activated cell sorting (FACS)-based techniques that can yield either a mixture or separate populations of undifferentiated and early differentiating spermatogonia. Once isolated, these cells may be useful for downstream analyses such as the examination of genes differentially regulated during self-renewal and early differentiation (Fig. 3).

1.1. Reporter Strain-Specific Differences

Choice of Pou5f1-GFP reporter strain is important, as several strains that have become available do not express GFP in postnatal or adult germ cells. The one strain that is successful for identifying a specific subpopulation of germ cells up to late adulthood (as demonstrated by our laboratory; Fig. 2e) is the transgenic strain B6;CBA-Tg(Pou5f1-EGFP)2Mnn/J (hereafter referred to as Pou5f1-GFP[Mann]). Originally reported in ref. 14, this strain is currently available through The Jackson Laboratory (Bar Harbor, ME). In comparison, we found that a knock-in reporter strain originally reported in ref. 15 – now available through The Jackson Laboratory as B6;129S4-Pou5f1tm2Jae/J – does not express GFP in postnatal testis. Furthermore, the transgenic strain originally reported in ref. 4, although generated similar to Pou5f1-GFP[Mann], reports Type A spermatogonia during the first wave of spermatogenesis – up to approximately 10 days postpartum (dpp) – but not during subsequent waves (4, 16). Although the mechanism underlying these strain-specific differences is uncertain, differences in the genomic location of the transgene and differential chromatin silencing of these regions may likely play a role, as has been reported for other reporters and transgenes (17–19).

1.2. In Vitro Use

The Pou5f1-GFP reporter has been used with much success for labeling induced pluripotent stem (iPS) cells and embryonic stem (ES) cells in culture (20, 21). GFP+ type A spermatogonia from Pou5f1-GFP[Mann] testes are detectable in situ and can be isolated from the other cells of the seminiferous epithelium by FACS (Fig. 2). However, they begin to lose GFP expression at the onset of culture. By 72 h in culture, the signal becomes undetectable, but cells continue to grow. Interestingly, GFP expression re-emerges after 3–4 weeks of culture. This observation was originally reported by Izadyar and colleagues (10) and confirmed in our laboratory using a variety of culture conditions supporting short- and long-term expansion of mouse SSCs (22–25). Therefore, the usefulness of this particular reporter strain for in vitro studies is limited in this respect.

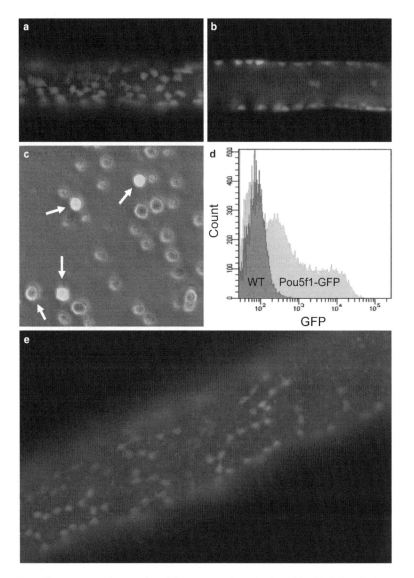

Fig. 2. Fluorescence micrograph and flow cytometric evaluation of freshly isolated tubules and cells isolated from Pou5f1-GFP^{Mann} mice. (**a, b**) Longitudinal cross-section adjacent to basement membrane (**a**) and through lumen (**b**) of a live, freshly isolated Pou5f1-GFP^{Mann} seminiferous tubule, demonstrating distribution of GFP+ spermatogonia along the basement membrane at 10 dpp. (**c**) GFP+ cells (*arrows*) in a freshly isolated, nonenriched primary cell population from 10 dpp Pou5f1-GFP^{Mann} mice. (**d**) Flow cytometry fluorescence intensity profile of wild-type (control) and Pou5f1-GFP^{Mann} freshly isolated cells from 10 dpp mice demonstrating a large population of GFP+ spermatogonia available for specific isolation through fluorescence-activated cell sorting (FACS). (**e**) Representative fluorescence micrograph of a freshly isolated tubule isolated from Pou5f1-GFP^{Mann} mice at 100 dpp demonstrating A_{aligned} spermatogonia.

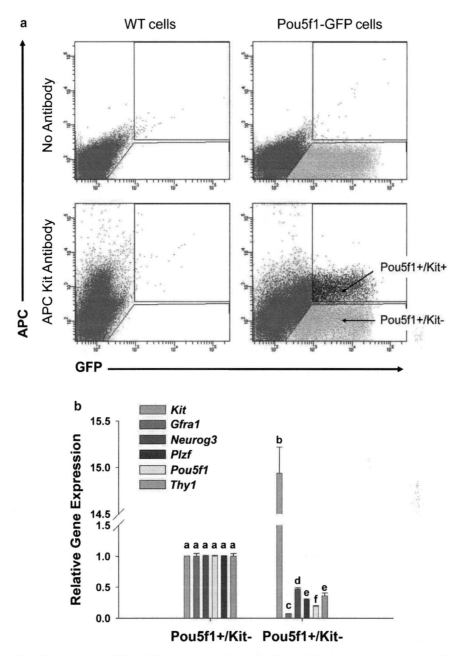

Fig. 3. Undifferentiated and early differentiating spermatogonia isolated through fluorescence-activated cell sorting (FACS). (a) Flow cytometric analysis and sorting of wild-type and Pou5f1-GFP^Mann cells stained with APC-Kit antibody. (b) Real-time-PCR analysis of Kit and canonical markers of undifferentiated spermatogonia in Pou5f1+/Kit– and Pou5f1+/Kit+ cells sorted from panel (a). Cells were processed for RT-PCR using the TaqMan Gene Expression Cells-to-CT Kit (Applied Biosystems, Carlsbad, CA) according to the manufacturer's instructions. The expression value of each gene was normalized to the amount of an internal control gene (Eif3l and Rps3) (29) and a relative quantitative fold change was determined using the $\Delta\Delta$Ct method. Experiments were performed in triplicate and data are represented as mean ± standard error. Taqman Gene Expression Assays (Applied Biosystems, Carlsbad, CA) used for specific transcripts were: Mm00460859_m1 (Eif3l), Mm00833897_m1 (Gfra1), Mm00445212_m1 (Kit), Mm00437606_s1 (Neurog3), Mm00658129_gH (Pou5f1), Mm00656272_m1 (Rps3), Mm00493681_m1 (Thy1), and Mm01176868_m1 (Zbtb16). Normalization to Eif3l and Rps3 produced identical results.

2. Materials

2.1. Mice

1. Pou5f1-GFP^Mann mice (7–10 dpp) as previously described (14) (The Jackson Laboratory) (see Note 1).

2. Age and strain matched nonreporter mice required for isolation of negative control cells devoid of GFP expression.

2.2. Buffers and Enzyme Solutions

1. Dulbecco's phosphate-buffered saline (DPBS) without Ca^{2+} and Mg^{2+} (Invitrogen).

2. Dulbecco's modified Eagle medium: Nutrient Mixture F-12 (DMEM/F12) (Invitrogen).

3. Hanks' balanced salt solution (HBSS) with Ca^{2+} and Mg^{2+} (Invitrogen).

4. HBSS without Ca^{2+} and Mg^{2+} (Invitrogen).

5. Cell dissociation buffer, enzyme free, Hanks'-based (Invitrogen).

6. Collagenase IV/DNase solution: Prepare 1 mg/ml Type IV collagenase (Sigma-Aldrich) and 2 μg/ml DNase (Sigma-Aldrich) in DMEM/F12. Filter-sterilize through a 0.22-μm filter, aliquot and freeze at –20 or –80°C. Thaw at room temperature immediately before use.

7. Trypsin/DNase solution: Dissolve DNase in 0.25% (w/v) trypsin with EDTA to a concentration of 200 μg/ml. Filter-sterilize through a 0.22 μm filter, aliquot and freeze at –20 or –80°C. Thaw at room temperature immediately before use.

8. S-HBSS buffer (supplemented HBSS buffer) (26): Prepare 20 mM HEPES, 6.6 mM sodium pyruvate, and 0.05% sodium lactate in HBSS with Ca^{2+} and Mg^{2+}. Filter-sterilize through a 0.22 μm filter and store at 4°C for up to 3 weeks.

9. Collagenase I solution: Dissolve Type I collagenase (Sigma-Aldrich) in S-HBSS to a concentration of 100 U/ml. Filter-sterilize through a 0.22-μm filter. Make fresh immediately before use.

10. FACS buffer (PBS-S) (27): Prepare 10 mM HEPES, 1 mM Pyruvate, 1 mg/ml glucose, 1% FBS (v/v), and 1× penicillin/streptomycin in 1× DPBS. Filter-sterilize through a 0.22 μm filter and store at 4°C for up to 3 weeks.

11. Trypan blue, 0.4% Solution (Sigma-Aldrich).

12. 4′,6-Diamidino-2-phenylindole (DAPI) stock solution: Dissolve 1 mg of DAPI (Sigma-Aldrich) in 100 ml of dH_2O. Protect from light and store at 4°C for up to 3 weeks. For long-term storage, aliquot and freeze at –20°C.

**2.3. Supplies
and Equipment**

1. Dissecting stereomicroscope with fiber-optic light source.
2. Dissection scissors.
3. Fine-tip forceps.
4. Class II, Type A2 HEPA-filtered biosafety cabinet.
5. Microcentrifuge, preferably with temperature control for maintenance at 4°C.
6. Benchtop centrifuge with buckets suitable for 15 and 50 ml conical tubes, preferably with temperature control for maintenance at 4°C.
7. Inverted fluorescence microscope (Olympus IX70, or similar) equipped with phase contrast objectives (phase $\times 4$, phase $\times 10$, phase $\times 20$, phase $\times 40$).
8. Cell counter or hemocytometer.
9. Small end-over-end rotator.
10. BD FACSAria II cell sorter (BD Biosciences), or similar, with BD FACSDiva software (BD Biosciences), or similar, for data acquisition.
11. CO_2 incubator with CO_2, humidity and temperature control, and monitoring.
12. Water Bath Shaker (New Brunswick Scientific C76, or similar).

**2.4. Plasticware/
Disposables**

1. Sterile 100 mm petri dishes.
2. Sterile 10 and 25 ml disposable pipettes.
3. 0.22 μm, 150 ml, Sterile disposable filter systems (Nalgene).
4. Sterile 15 and 50 ml polypropylene conical tubes.
5. Sterile 1.5 ml polypropylene microcentrifuge tubes.
6. 40 and 70 μm Nylon mesh cell strainers (BD Biosciences).
7. 5 ml (12×75 mm) Polypropylene round bottom FACS collection tubes. (BD Biosciences).
8. 5 ml (12×75 mm) Polypropylene round bottom FACS collection tube with 35 μm nylon mesh cell strainer cap (BD Biosciences).

3. Methods

Single cell suspensions of mouse testicular cells enriched for seminiferous epithelium cells (germ cells and Sertoli cells) are isolated from a pool of Pou5f1-GFP[Mann] pups that are 7–10 days postpartum (see Note 2). Four to eight pups are used per isolation, although a greater number may be necessary, in which case the volumes and vessels used should be scaled up proportionally. Using four to eight pups, the yield will range from 1 to 2.5×10^6 cells/pup; 4 to 20×10^6 cells total.

3.1. Basic Germ Cell Isolation

3.1.1. Seminiferous Tubule Isolation

1. Sacrifice pups, aseptically excise testes with adjoining epidydimis, and pool into one 100 mm Petri dish containing chilled DMEM/F-12. If a large number of testes are collected and/or an accumulation of blood and debris is present, tissues may be transferred individually with forceps to a new 100 mm Petri dish containing fresh DMEM/F-12 to wash.

2. With a fine-tipped forcep in one hand, secure one testis by the efferent ducts, and with a second fine-tipped forcep in closed position in the other hand, puncture the tunica albuginea. With the tips of the closed forcep still inside the testis, allow the tips to spread apart, tearing the tunica albuginea open and exposing the seminiferous tubules. Once the seminiferous tubules are sufficiently exposed, in a second continuous motion using the same forcep utilized to puncture/spread, extract the seminiferous tubules, which with practice can typically be done in entirety with one intact clump.

3. Transfer seminiferous tubule clumps into one 15 ml conical tube on ice containing 5–10 ml of fresh DMEM/F-12 (see Note 3). Pool tubule clumps from age-matched nonreporter pups (flow cytometry negative control for GFP expression) into a separate tube and process in parallel.

3.1.2. Sequential Digestion: Concept

For the isolation of seminiferous epithelium cells and concomitant depletion of interstitial cells (blood cells, peritubular myoid cells, and Leydig cells), a two-step digestion protocol is used. The first gentle collagenase digestion breaks the aggregated tubules and tubule clumps into individual tubule fragments, and releases the interstitial cells into solution. Tubule fragments are allowed to settle at 1 U gravity, and the interstitial cells, which will remain in the supernatant, are removed. Gravity sedimentation is repeated again during a wash step and the cleaned up tubule fragments are resuspended and incubated in trypsin solution or cell dissociation buffer with mechanical disruption. After straining to remove any undissociated cells and agglomerated cellular debris, the resulting single cell suspension can be processed for FACS.

3.1.3. Tubule Fragmentation and Cleanup (FACS Isolation of Pou5f1+ Cells Only)

1. Ensure all tubule clumps have settled to the bottom of the tube.

2. Suction/pipette off DMEM/F-12, add 4–8 ml freshly thawed Collagenase IV solution, and pipette up and down gently several times (see Note 4).

3. Incubate at room temperature for 1–2 min, pipette up and down gently several times again, and observe the status of the tubules. At this point, tubules should dissociate into tubule fragments. Repeat this step only if tubule clumps have not become dissociated. If only a small tubule clump remains, the following steps will complete the fragmentation process.

4. Allow the tubules to sediment, which can take as little as 2 min, but should not exceed 5 min to ensure the single dissociated interstitial cells remain in the supernatant.

5. Next resuspend the settled tubule fragments in an equal volume of DPBS without Ca^{2+} or Mg^{2+}, and pipette up and down gently (or invert if tubules are fragmented) and allow the tubule fragments to sediment again as described above (see Note 5).

3.1.4. Trypsin Digestion (for FACS Isolation of Pou5f1+ Cells)

1. Once the tubule fragments have sedimented, remove DPBS, and resuspend tubule fragments in an equal volume of trypsin solution and incubate at room temperature with intermittent pipetting (see Note 6).

2. Total incubation time in trypsin solution – to yield a single cell suspension with only few small intact tubule fragments remaining – ranges from 5 to 10 min, but can take longer if the activity of trypsin in low (see Note 7).

3. Digestion with intermittent pipetting is continued until completed and followed by addition of FBS to a final concentration of 10%, to quench trypsin activity. Subsequent steps are carried out at 4°C.

4. The cells are centrifuged at $300 \times g$ for 7 min, then washed and resuspended in FACS buffer.

5. Passage of cells through a 70 and/or 40 μm strainer to obtain a single cell suspension can take place any step after trypsin has been inactivated (see Note 8).

6. Count cells using a viability dye such as trypan blue to determine viable and nonviable cell numbers.

3.2. Modifications of Basic Germ Cell Isolation for Preservation of Cell Surface Antigens (for FACS Isolation of Pou5f1+/Kit– and Pou5f1+/Kit+ Cells)

Since trypsin will act proteolytically upon cell surface proteins, to ensure preservation of cell surface antigens and optimal staining of Kit receptor, the nontrypsin-based isolation protocol of Vincent et al. is used (26).

3.2.1. Tubule Fragmentation and Cleanup (FACS Isolation of Pou5f1+ Cells Only)

1. Instead of Collagenase IV solution, use freshly prepared Collagenase I solution and extend the incubation conditions from approximately 5 min at room temperature, with occasional pipetting, to 20–30 min in a reciprocating water bath at 32°C (see Note 9).

2. Pellet tubule fragments through gravity sedimentation, extending incubation to 10 min if necessary to ensure small tubule fragments have sedimented. If a fluorescence microscope is nearby during the isolation, microscopic evaluation of the suspension before and after sedimentation is helpful in confirming that the single cells remaining in suspension are depleted of GFP+ cells. If too many GFP+ cells were released during Collagenase I digestion, proceed to centrifuging the entire contents at $300 \times g$ for 7 min to ensure full recovery. Remove supernatant.

3. Resuspend pellet in HBSS without Ca^{2+} and Mg^{2+} to wash and centrifuge at $300 \times g$ for 7 min.

4. Resuspend pellet in cell dissociation buffer and incubate for 20–30 min at 32°C in a reciprocating water bath set at slow speed.

5. Pipette up and down to dissociate the cells further, and observe the cell suspension microscopically to determine whether a large quantity of single cells has been released. If not, incubate further with additional pipetting, then filter through a 40 μm nylon mesh to remove cell clumps (see Note 10).

6. Count cells using a viability dye such as trypan blue to determine viable and nonviable cell numbers.

3.2.2. Kit Receptor Staining

1. Once the cells are harvested according to the methods of Vincent et al., the cell suspension is adjusted to a concentration of 1×10^7 cells/ml in FACS buffer (see Notes 11 and 12).

2. Add allophycocyanin (APC) conjugated anti-mouse CD117 (Kit) antibody (BD Biosciences) to cells at a final concentration of 10 μg/ml and incubate with slow end-over-end rotation at 4°C for 30 min in the dark (see Note 13).

3. Wash the cells three times with FACS buffer (wash volume same as incubation volume) and resuspend the final cell suspension again at a concentration of 1×10^7 cells/ml in FACS buffer.

4. The cells should be protected from light on ice or in a fridge until they are processed (see Note 14).

3.2.3. Preparation of FACS Collection Tubes

1. The day before cell sorting, fill FACS collection tubes entirely with DMEM/F12 + 10% FBS and store in a refrigerator overnight to precoat tubes (see Note 15).

2. Transfer prepared cells and DMEM/F12 + 10% FBS filled tubes on ice to FACS instrument.

3. Immediately preceding sort, remove approximately 50% volume of DMEM/F12 + 10% FBS used to precoat tubes and collect instrument-sorted cells into precoated FACS collection tubes kept chilled at 4°C during sort.

3.2.4. FACS Controls

Autofluorescence and fluorescence spillover are eliminated during FACS through proper set-up and initial compensation of the flow cytometer; thorough use of no fluorophore and single fluorophore controls prior to each run is imperative. Testicular cells simultaneously isolated from age-matched wild-type mice (as mentioned above) are run through the instrument first to set the cutoff value for background fluorescence. Unstained Pou5f1-GFP cells and APC-Kit antibody stained wild-type cells are used for compensating between the multiple color channels.

3.2.5. FACS Instrument

In our studies, the instrument used was the BD FACS Aria II cell sorter. With this instrument GFP is excited with a blue 488 nm solid state laser and emission light is detected through a 530/30 nm bandpass filter; APC is excited with a red 640 nm solid state laser and emission light is detected through a 670/30 nm bandpass filter. The BD FACSDiva software is used to control the instrument before and during data acquisition and for data analysis during and after the run. Samples are sorted at 4°C using an 85 μm nozzle and 45 psi pressure.

3.2.6. Sorting

1. Immediately prior to processing samples, add 5 μl DAPI stock solution per 1 ml cells (see Note 16).

2. Gate for DAPI negative cells to exclude dead cells (see Note 16).

3. Gate for singlets on a plot of forward scatter area (FSC-A) versus forward scatter width (FSC-W) to ensure ultimate purity.

4. For sorting GFP+ cells only, gate for GFP+ cells on a plot of GFP versus FSC-A, using wild-type cells to determine the cutoff value for background fluorescence. For sorting Pou5f1+/Kit− and Pou5f1+/Kit+ cells separately, gate for each population on a plot of GFP versus APC intensity (see Fig. 3a).

5. Sort cells/events from gated regions of interest into precoated tubes (as mentioned above). Maintain tubes at 4°C or on ice during and after the sort.

When isolating cells from 8 dpp mice, one can expect to retrieve approximately 20% GFP+ cells; if stained for c-Kit, roughly two-thirds will be Pou5f1+/Kit− and one-third Pou5f1+/Kit+.

4. Notes

1. All experiments involving live mice must conform to national and institutional regulations.

2. This age is selected so as to enrich for premeiotic germ cells, in particular, Type A spermatogonia, and to obtain the highest overall yield possible (28).

3. As a rule of thumb, use 1 ml per two to four testes. Therefore, for a small-scale isolation from one to two pups, one 1.5 ml Eppendorf tube may be used for all steps. Otherwise, isolation from 8 pups (16 testes) will require at least 4 ml solution in a 15 ml conical tube; isolation from greater than 12 pups should be carried out in a 50 ml conical tube.

4. Use the lower volume of 4 ml if isolating from 7-day-old mice, and the higher volume of 8 ml if isolating from 10 day olds.

5. DPBS without Ca^{2+} or Mg^{2+} is used since the testis isolation and collagenase digestion was performed in DMEM/F12 to ensure optimal health, however, contains Ca^{2+} and Mg^{2+}, which promotes cell-to-cell adhesion.

6. Assuming the activity of trypsin is good, it is better to perform the trypsin digestion at room temperature rather than 37°C, as this allows one to make several separate microscopic observations of the status of digestion using a nearby microscope, with a lower likelihood of over digestion.

7. If trypsin activity is low, trypsin activity may be increased by placing the tube in a 37°C bath (without shaking) to warm.

8. If very little insoluble debris is present, to obtain the most cells, it is preferable to strain after the last wash, since cells may continue to dissociate during washes. However, if a significant amount of insoluble debris is observed after the trypsin digestion (if some overdigestion has taken place), it may be desirable to strain before the first centrifugation, with a second straining step after the last wash.

9. Collagenase I is used instead of Collagenase IV since it is gentler, yet allows for increased fragmentation and dissociation of the basement membrane, facilitating subsequent nonenzymatic digestion.

10. With the cell dissociation buffer protocol, complete digestion of all tubule fragments into single cells is less efficient than with trypsin, and a yield closer to $0.5–1 \times 10^6$ cells/pup should be expected; however, cell surface antigenicity will be retained. Scale-up accordingly if a greater number of cells are required.

11. Cells are stained in 1.5 ml Eppendorf tubes, although any suitable container may be used.

12. All staining procedures should be carried out with ice-cold reagents and on wet ice or at 4°C to suppress gene expression changes.

13. Since antibody binding efficiency and fluorescence intensity may vary based on lot and investigator handling, an initial antibody titration should be performed to obtain optimal results.

14. For best results, sort the cells as soon as possible, but the cells can be maintained in FACS buffer for several hours.

15. If collection tubes are uncoated, cells will adhere to the inside of the tubes, making recovery difficult and significantly decreasing yield.

16. Dead cell exclusion using DAPI may be excluded during the sort if cell viability was 100% as determined during cell counting with trypan blue.

Acknowledgments

The authors would like to thank Dr. Barbara Pilas and Ben Montez from the R. J. Carver Biotechnology Center at the University of Illinois at Urbana-Champaign for their invaluable assistance with flow cytometry and comments on the manuscript. The authors are also grateful to the NIH for funding this work.

References

1. Nicholas CR, Xu EY, Banani SF, Hammer RE, Hamra FK, and Reijo Pera RA (2009) Characterization of a Dazl-GFP germ cell-specific reporter. Genesis 47: 74–84

2. Yoshida S, Takakura A, Ohbo K, Abe K, Wakabayashi J, Yamamoto M, Suda T, and Nabeshima Y (2004) Neurogenin3 delineates the earliest stages of spermatogenesis in the mouse testis. Dev Biol 269: 447–458

3. Nayernia K, Li M, Jaroszynski L, Khusainov R, Wulf G, Schwandt I, Korabiowska M, Michelmann HW, Meinhardt A, and Engel W (2004) Stem cell based therapeutical approach of male infertility by teratocarcinoma derived germ cells. Hum Mol Genet 13: 1451–1460

4. Ohbo K, Yoshida S, Ohmura M, Ohneda O, Ogawa T, Tsuchiya H, Kuwana T, Kehler J, Abe K, Scholer HR, and Suda T (2003) Identification and characterization of stem cells in prepubertal spermatogenesis in mice small star, filled. Dev Biol 258: 209–225

5. Yeom YI, Fuhrmann G, Ovitt CE, Brehm A, Ohbo K, Gross M, Hubner K, and Scholer HR (1996) Germline regulatory element of Oct-4 specific for the totipotent cycle of embryonal cells. Development 122: 881–894

6. Yoshimizu T, Sugiyama N, De Felice M, Yeom YI, Ohbo K, Masuko K, Obinata M, Abe K, Scholer HR, and Matsui Y (1999) Germline-specific expression of the Oct-4/green fluorescent protein (GFP) transgene in mice. Dev Growth Differ 41: 675–684

7. Pesce M, Wang X, Wolgemuth DJ, and Scholer H (1998) Differential expression of the Oct-4 transcription factor during mouse germ cell differentiation. Mech Dev 71: 89–98

8. Tadokoro Y, Yomogida K, Ohta H, Tohda A, and Nishimune Y (2002) Homeostatic regulation of germinal stem cell proliferation by the GDNF/FSH pathway. Mech Dev 113: 29–39

9. Schrans-Stassen BH, van de Kant HJ, de Rooij DG, and van Pelt AM (1999) Differential expression of c-kit in mouse undifferentiated and differentiating type A spermatogonia. Endocrinology 140: 5894–5900

10. Izadyar F, Pau F, Marh J, Slepko N, Wang T, Gonzalez R, Ramos T, Howerton K, Sayre C, and Silva F (2008) Generation of multipotent cell lines from a distinct population of male germ line stem cells. Reproduction 135: 771–784

11. Shinohara T, Orwig KE, Avarbock MR, and Brinster RL (2000) Spermatogonial stem cell enrichment by multiparameter selection of mouse testis cells. Proc Natl Acad Sci USA 97: 8346–8351

12. Boiani M, Kehler J, and Scholer HR (2004) Activity of the germline-specific Oct4-GFP transgene in normal and clone mouse embryos. Methods Mol Biol 254: 1–34

13. Hammond SS, and Matin A (2009) Tools for the genetic analysis of germ cells. Genesis 47: 617–627

14. Szabo PE, Hubner K, Scholer H, and Mann JR (2002) Allele-specific expression of imprinted genes in mouse migratory primordial germ cells. Mech Dev 115: 157–160

15. Lengner CJ, Camargo FD, Hochedlinger K, Welstead GG, Zaidi S, Gokhale S, Scholer HR, Tomilin A, and Jaenisch R (2007) Oct4 expression is not required for mouse somatic stem cell self-renewal. Cell Stem Cell 1: 403–415

16. Ohmura M, Yoshida S, Ide Y, Nagamatsu G, Suda T, and Ohbo K (2004) Spatial analysis of germ stem cell development in Oct-4/EGFP transgenic mice. Arch Histol Cytol 67: 285–296

17. Garrick D, Fiering S, Martin DI, and Whitelaw E (1998) Repeat-induced gene silencing in mammals. Nat Genet 18: 56–59

18. Garrick D, Sutherland H, Robertson G, and Whitelaw E (1996) Variegated expression of a globin transgene correlates with chromatin accessibility but not methylation status. Nucleic Acids Res 24: 4902–4909

19. Henikoff S (1998) Conspiracy of silence among repeated transgenes. Bioessays 20: 532–535

20. Meissner A, Wernig M, and Jaenisch R (2007) Direct reprogramming of genetically unmodified fibroblasts into pluripotent stem cells. Nat Biotechnol 25: 1177–1181

21. Chen S, Do JT, Zhang Q, Yao S, Yan F, Peters EC, Scholer HR, Schultz PG, and Ding S (2006) Self-renewal of embryonic stem cells by a small molecule. Proc Natl Acad Sci USA 103: 17266–17271

22. Guan K, Wolf F, Becker A, Engel W, Nayernia K, and Hasenfuss G (2009) Isolation and cultivation of stem cells from adult mouse testes. Nat Protoc 4: 143–154

23. Hofmann MC, Braydich-Stolle L, and Dym M (2005) Isolation of male germ-line stem cells; influence of GDNF. Dev Biol 279: 114–124

24. Kanatsu-Shinohara M, Ogonuki N, Inoue K, Miki H, Ogura A, Toyokuni S, and Shinohara T (2003) Long-term proliferation in culture and germline transmission of mouse male germline stem cells. Biol Reprod 69: 612–616

25. Kubota H, Avarbock MR, and Brinster RL (2004) Growth factors essential for self-renewal and expansion of mouse spermatogonial stem cells. Proc Natl Acad Sci USA 101: 16489–16494

26. Vincent S, Segretain D, Nishikawa S, Nishikawa SI, Sage J, Cuzin F, and Rassoulzadegan M (1998) Stage-specific expression of the Kit receptor and its ligand (KL) during male gametogenesis in the mouse: a Kit-KL interaction critical for meiosis. Development 125: 4585–4593

27. Kubota H, Avarbock MR, and Brinster RL (2004) Culture conditions and single growth factors affect fate determination of mouse spermatogonial stem cells. Biol Reprod 71: 722–731

28. Bellve AR, Cavicchia JC, Millette CF, O'Brien DA, Bhatnagar YM, and Dym M (1977) Spermatogenic cells of the prepuberal mouse. Isolation and morphological characterization. J Cell Biol 74: 68–85

29. de Jonge HJ, Fehrmann RS, de Bont ES, Hofstra RM, Gerbens F, Kamps WA, de Vries EG, van der Zee AG, te Meerman GJ, and ter Elst A (2007) Evidence based selection of housekeeping genes. PLoS One 2: e898

Chapter 4

Isolation of Human Male Germ-Line Stem Cells Using Enzymatic Digestion and Magnetic-Activated Cell Sorting

Zuping He, Maria Kokkinaki, Jiji Jiang, Wenxian Zeng, Ina Dobrinski, and Martin Dym

Abstract

Mammalian spermatogenesis is a process whereby male germ-line stem cells (spermatogonial stem cells) divide and differentiate into sperm. Although a great deal of progress has been made in the isolation and characterization of spermatogonial stem cells (SSCs) in rodents, little is known about human SSCs. We have recently isolated human G protein-coupled receptor 125 (GPR125)-positive spermatogonia and GDNF family receptor alpha 1 (GFRA1)-positive spermatogonia using a 2-step enzymatic digestion and magnetic-activated cell sorting (MACS) from adult human testes. Cell purities of isolated human GPR125- and GFRA1-positive spermatogonia after MACS are greater than 95%, and cell viability is over 96%. The isolated GPR125- and GFRA1-positive spermatogonia coexpress GPR125, integrin, alpha 6 (ITGA6), THY1 (also known as CD90), GFRA1, and ubiquitin carboxyl-terminal esterase L1 (UCHL1), markers for rodent or pig SSCs/progenitors, suggesting that GPR125- and GFRA1-positive spermatogonia are phenotypically the SSCs in human testis. Human GPR125-positive spermatogonia can be cultured for 2 weeks with a remarkable increase in cell number. Immunocytochemistry further reveals that GPR125-positive spermatogonia can be maintained in an undifferentiated state in vitro. Collectively, the methods using enzymatic digestion and MACS can efficiently isolate and purify SSCs from adult human testis with consistent and high quality. The ability of isolating and characterizing human SSCs could provide a population of stem cells with high purity for mechanistic studies on human SSC self-renewal and differentiation as well as potential applications of human SSCs in regenerative medicine.

Key words: Human male germ-line stem cells, Isolation, Enzymatic digestion, Magnetic-activated cell sorting, GPR125, GFRA1, Characterization

1. Introduction

Spermatogenesis is a process that involves the proliferation and differentiation of male germ-line stem cells (spermatogonial stem cells) into sperm in mammals. Studies on spermatogonial stem cells (SSCs) are of particular significance in view of their unique

Wai-Yee Chan and Le Ann Blomberg (eds.), *Germline Development: Methods and Protocols*, Methods in Molecular Biology, vol. 825, DOI 10.1007/978-1-61779-436-0_4, © Springer Science+Business Media, LLC 2012

characteristics: (1) SSCs are the stem cells that transmit genetic information from one generation to subsequent generations (1); (2) SSCs may soon be differentiated into sperm in vitro to help infertility patients via the In Vitro *Fertilization* and *Embryo Transfer (IVF-ET)*. Around 15% of couples are infertile around the world, and half can be attributed to male infertility. Clinicians often take a testicular biopsy and try to isolate cells that can be used for *Intra Cytoplasmic Sperm Injection (ICSI)*. In vitro germ cell differentiation from SSCs has the potential to help infertile men father children. (3) There is the potential use of SSC transplantation (2, 3) to restore fertility in cancer patients after chemotherapy or irradiation therapy as reviewed in refs. 3, 4. (4) SSCs could be used as a target for a male contraceptive. It is possible to specifically target the SSCs with small molecules (e.g., microRNA inhibitors) to selectively interfere with SSC differentiation, thus leading to a novel male contraceptive. (5) SSCs are an excellent model to elucidate the mechanisms controlling stem cell renewal versus differentiation (5), since SSCs can self-renew throughout their lifetime and they can differentiate into spermatocytes, spermatids, and eventually sperm. Lastly and most significantly, numerous studies have recently demonstrated that mouse and human SSCs can acquire pluripotency to become embryonic stem (ES)-like cells that can differentiate into all cell lineages of the three germ cell layers (6–14). This provides new and promising therapeutic prospects for human SSCs to generate various mature cells for regenerative medicine and treatment of human diseases without ethical issues associated with human ES cells (15, 16).

There is a great deal of information available on the biochemical characterization and isolation of SSCs in rodents and other species. We and others have demonstrated that the GDNF family receptor alpha 1 (GFRA1) is a surface marker for SSCs and progenitor cells in the mouse testis (17–20), while others also showed integrin alpha 6 (ITGA6) and THY1 (also known as CD90) to be surface markers for mouse SSCs and progenitor cells (21, 22). Ubiquitin carboxyl-terminal hydrolase L1 (UCHL1) is regarded as a marker for porcine, bovine, and nonhuman primate spermatogonia (23). Of note, we have demonstrated, using immunohistochemistry, that UCHL1 is also expressed in human spermatogonia (24). Recently, G protein-coupled receptor 125 (GPR125) has been shown to be a marker for mouse SSCs and their progeny (8). In addition, SSCs can be separated efficiently from rodent testis and cultured for short and long periods (18, 25, 26), which makes it feasible to examine molecular mechanisms that regulate fate decisions of rodent SSCs.

In the human testis, Clermont first distinguished A_{dark} and A_{pale} spermatogonia and speculated that A_{dark} spermatogonia were the reserve stem cells and A_{pale} spermatogonia were the renewing stem cells (27–30). However, information on the isolation and characterization of human SSCs is limited (31). One major reason for little progress on human SSCs has been the difficulty in gaining access to sufficient quantities of normal human testis for research purposes. We have identified a novel source of human testicular material, namely, from recently (~1–2 h from death) deceased organ donors. Recently, we revealed that GPR125 and GFRA1 are expressed in a subset of human spermatogonia as shown by immunohistochemistry (24). Magnetic-activated cell sorting (MACS) technology has been used to effectively separate SSCs from mouse testis (32, 33). We were the first to report the isolation of GPR125-positive spermatogonia using a 2-step enzymatic digestion and MACS from adult human testes (24). In the sections below, we provide detailed descriptions of the procedures to effectively isolate and characterize human GPR125- and GFRA1-positive spermatogonia: (1) the 2-step enzymatic digestion (Fig. 1a, b), (2) MACS (Fig. 1c), and (3) immunocytochemical analysis of the freshly isolated GPR125- and GFRA1-positive spermatogonia (Figs. 1d–p and 2) as well as GFRA1-negative male germ cells (Fig. 3). Using these approaches, it is feasible to obtain human SSCs with the highest purity and viability for further studies on molecular mechanisms regulating human SSC self-renewal and differentiation as well as for potentially important implications of SSCs in regenerative medicine.

2. Materials

2.1. Procurement of Testes from Organ Donors

1. Testes can be obtained from organ donors to the Washington Regional Transplant Consortium (Annandale, VA).
2. Heparin.
3. Viaspan™.
4. Dulbecco's modified Eagle's medium (DMEM) containing 1 × antibiotic.
5. Freezing solution: 10% DMSO, 80% fetal bovine serum (FBS), and 10% DMEM.

2.2. Enzymatic Digestion of Human Testis Tissues

1. Enzymatic digestion solution I: 15 ml DMEM/F12 with 2 mg/ml collagenase IV or collagenase XI and 1 μg/μl DNase I.
2. Enzymatic digestion solution II: 15 ml DMEM/F12 with 4 mg/ml collagenase IV or collagenase XI, 2.5 mg/ml hyaluronidase, 2 mg/ml trypsin, and 1 μg/μl DNase I.
3. DMEM/F12 medium supplemented with 10% FBS.
4. NU serum.
5. Sterile acrodisc syringe filter (5 μm).

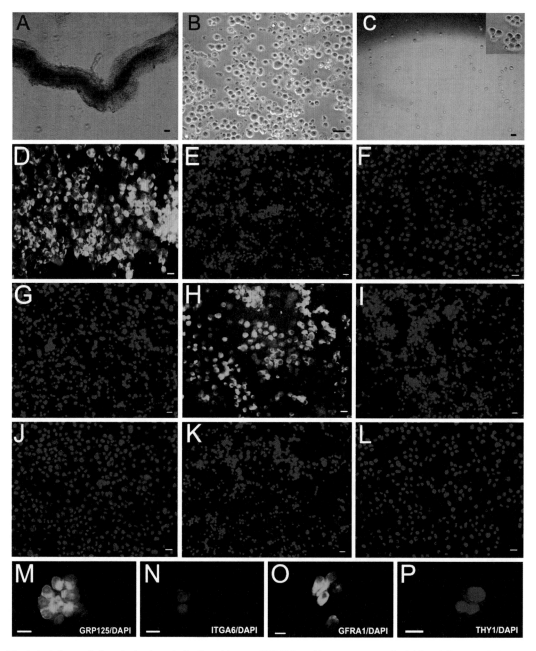

Fig. 1. Isolation and phenotypic characterization of human GPR125-positive spermatogonia. (a) Seminiferous tubules are isolated from human testes by mechanical dissociation and the first enzymatic digestion. (b) Germ cells are obtained by a second enzymatic digestion and differential plating. (c) GPR125-positive spermatogonia are obtained by MACS with an antibody to GPR125. The photo insert in c is a high magnification view showing GPR125-positive spermatogonia. Immunocytochemistry shows the expression of GPR125 (d), ITGA6 (f), GFRA1 (h), and THY1 (j), in freshly isolated GPR125-positive spermatogonia. (m–p) High magnification image shows that freshly isolated GPR125-positive spermatogonia express GPR125 (m), ITGA6 (N), GFRA1 (o), and THY1 (p). No staining is observed in GPR125-negative male germ cells using antibody to GPR125 (e), ITGA6 (g), GFRA1 (i), or THY1 (k). (l) Replacement of primary antibody with PBS in the isolated GPR125-positive spermatogonia serves as a negative control. Scale bars in (a, c, d, f, h, j, l–o), and $P=10$ μm; bars in (b, e, g, i), and $K=20$ μm (Adapted from ref. 24. ©Society for the Study of Reproduction).

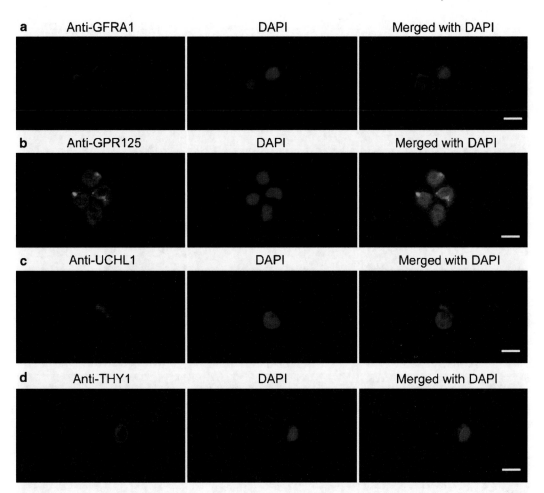

Fig. 2. Phenotypic characterization of the freshly isolated GFRA1-positive spermatogonia from adult human testis. After MACS, immunocytochemical analysis showed the expression of GFRA1 (**a**), GPR125 (**b**), UCHL1 (**c**), and THY1 (**d**), in isolated GFRA1-positive spermatogonia. DAPI was used to stain cell nuclei. Scale bars in **a–d** = 10 μm.

2.3. Differential Plating to Obtain Human Male Germ Cells

2.4. Isolation of Human GPR125- and GFRA1-Positive Spermatogonia by MACS

Culture medium: DMEM/F12 supplemented with 10% FBS; 0.1% gelatin.

1. DMEM supplemented with 10% NU serum and 1 μg/μl DNase I.

2. Rabbit polyclonal antibody to GPR125 (Abcam Inc.).

3. Rabbit polyclonal antibody to GFRA1 (Santa Cruz Biotechnology, Inc).

4. Trypan blue solution (0.4%).

5. BSA–EDTA–PBS buffer: ice-cold PBS containing 0.5% BSA and 2 mM EDTA.

6. MACS BSA stock solution (Miltenyi Biotec).

7. MACS Preseparation filters (Miltenyi Biotec).

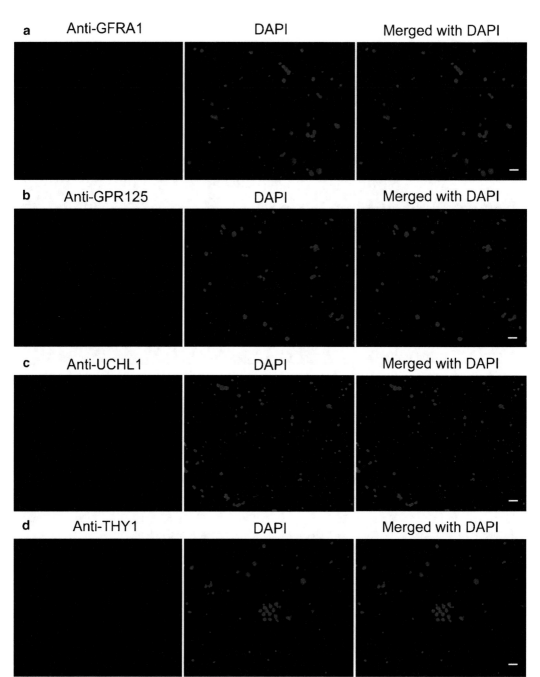

Fig. 3. Phenotypic characterization of the freshly isolated GFRA1-negative male germ cells from adult human testis. After MACS, immunocytochemical analysis showed the expression of GFRA1 (**a**), GPR125 (**b**), UCHL1 (**c**), and THY1 (**d**), in the freshly isolated GFRA1-negative male germ cells. DAPI was used to stain cell nuclei. Scale bars in **a–d** = 10 μm.

8. MACS Separation Columns (Miltenyi Biotec).

9. MACS Separator (Miltenyi Biotec).

10. Goat anti-rabbit IgG Microbeads (Miltenyi Biotec).

2.5. Immunocyto-chemistry of Isolated Human GPR125- and GFRA1-Positive Spermatogonia

1. 4% Paraformaldehyde (37% of paraformaldehyde is diluted with phosphate-buffered saline (PBS) to final concentration of 4%).

2. Blocking solution: 10% goat serum (Invitrogen).

3. Rabbit polyclonal antibody to GPR125 (2 μg/ml in PBS).

4. Rabbit polyclonal antibody to GFRA1 (2 μg/ml in PBS).

5. Mouse monoclonal antibody to THY1 (LifeSpan Biosciences) (1 μg/ml in PBS).

6. Rabbit polyclonal antibody to UCHL1 (AbD Serotec) (1 μg/ml in PBS).

7. Phycoerythrin (PE)-conjugated mouse monoclonal antibody to ITGA6 (BD Biosciences) (1 μg/ml in PBS).

8. Normal rabbit IgG (Santa Cruz Biotechnology Inc.) (2 μg/ml in PBS).

9. Normal mouse IgG (Santa Cruz Biotechnology Inc.) (2 μg/ml in PBS).

10. Fluorescein (FITC)-conjugated goat anti-rabbit (Jackson ImmunoResearch Laboratories) at a 1:200 dilution in PBS.

11. Rhodamine-conjugated goat anti-rabbit or anti-mouse IgG (Jackson ImmunoResearch Laboratories) at a 1:200 dilution in PBS.

12. DAPI (4′, 6′-diamidino-2-phenylindole).

3. Methods

3.1. Procurement of Testes from Organ Donors

During organ retrieval in a hospital operating suite, 30,000 U of heparin is injected intravenously prior to aortic "cross clamping" in the thorax and in the lower abdomen (24). After clamping, the aorta is immediately perfused with Viaspan™ (see Note 1), a standard organ preservation solution. The testes are retrieved after a 10-min perfusion, packed in cold Viaspan™, and sent to the laboratory within 1–2 h of removal from the donor. The testes are immediately placed aseptically in DMEM containing 1×antibiotic. The testis tissue is frozen as ~1 g pieces in freezing solution, including 10% DMSO, 80% FBS, and 10% DMEM, in liquid nitrogen, and the integrity of the tissue should be excellent.

3.2. Enzymatic Digestion of Human Testis Tissue

1. The UV in the hood is turned on at least 30 min prior to use.

2. Enzymes including collagenase IV or collagenase XI (4°C), hyaluronidase, trypsin, and DNase I (kept at –20°C) are taken from the refrigerator.

3. Thirty milligrams of collagenase IV or collagenase XI is weighed, put in a 15-ml conical tube, and kept at 4°C.

4. Sixty milligrams of collagenase IV or collagenase XI, 37.5 mg of hyaluronidase, and 30 mg of trypsin are weighed, put in a 15-ml conical tube, and placed at 4°C.

5. The enzymes are returned to the refrigerator at 4°C immediately after the weighing. DNase I is thawed on wet ice.

6. The donor or frozen human testicular tissues are placed in 100 ml of DMEM/F12 in 120 ml container. All subsequent procedures are carried out in the sterile hood.

7. After weighing, testicular tissues (~1 or 2 g) are placed in 10 ml of DMEM/F12 and cut into small pieces measuring approximately $1 \times 1 \times 1$ mm and washed three times in DMEM/F12 to eliminate any remaining blood cells.

8. Testicular tissues are minced using scissors until a semiliquid state has been achieved. After the addition of 20 ml of DMEM/F12, the tissues are transferred to a new 120-ml container.

9. 20 ml of fresh DMEM/F12 is added to the container.

10. Enzyme I is taken out of the refrigerator (step 3) and dissolved in 15 ml of DMEM/F12 to yield 2 mg/ml collagenase IV or collagenase XI.

11. The dissolved collagenase IV or collagenase XI is sterilized in a 10-ml syringe mounted with a Sterile Acrodisc Syringe filter (5 μm). DNase I is added to enzyme I to yield 1 μg/μl DNase I and mixed gently with a pipette; this is enzyme solution I.

12. The minced testis tubules from step 8 are transferred to a 50-ml conical tube using a 10 ml glass pipette (plastic pipette is not suitable, see Note 2) and fresh DMEM/F12 is added to get a final volume of 50 ml. After a 5-min sedimentation, the supernatant is removed with a 25-ml pipette.

13. Enzyme solution I (from step 11) is added to the minced testis tissues (step 12) and incubated in a shaking water bath at 34°C for about 10 or 15 min.

14. The digested testis tissues are examined under a microscope to make sure that there are only seminiferous tubules with no gross lumps (see Note 3). If lumps remain, the testis tissue is pipetted up and down with a 5-ml glass pipette (plastic pipette is not suitable, see Note 2) for 1 min and incubated again for another 5 min until only seminiferous tubules are obtained. Almost all the interstitial tissue clumps should be removed after this procedure.

15. The solution containing seminiferous tubules is transferred to a new 50-ml conical tube and 30 ml of fresh DMEM/F12 containing 10% FBS is added to stop the digestion. After a 5-min sedimentation, the supernatant is removed with a 25-ml pipette and 50 ml of fresh DMEM/F12 is added to the pellet to wash the seminiferous tubules.

16. The seminiferous tubules are washed extensively to remove all Leydig cells and other interstitial cells: DMEM/F12 is added to the conical tube and sedimented for 5 min; the supernatant is removed and the wash is repeated another three times (see Note 4).

17. The enzyme mixture (from step 4) is taken out of the refrigerator and dissolved in 15 ml of DMEM/F12 to yield 4 mg/ml collagenase IV or collagenase XI, 2.5 mg/ml hyaluronidase, and 2 mg/ml trypsin.

18. The enzyme mixture is sterilized in a 10-ml syringe mounted with a Sterile Acrodisc Syringe filter (5 μm). DNase I is added to yield 1 μg/μl DNase I and mixed gently with 5-ml pipette. This is enzyme solution II.

19. The supernatant containing the seminiferous tubules (from step 16) is removed with a 25-ml pipette, and enzyme solution II is added to the tubules.

20. The mixture of tubules and enzyme solution II is transferred to a fresh container using a 10-ml glass pipette (plastic pipette is not suitable, see Note 2) and incubated in a shaking bath for 15 min at 100 cycles/min.

21. After a 15 min enzyme digestion, pipette the mixture gently 10–15 times using a 10-ml glass pipette (plastic pipette is not suitable, see Note 2) without producing air bubbles.

22. A drop of the mixture is placed on a microscope slide and examined with an inverted microscope to assess the effectiveness of the second enzyme digestion (see Note 5).

23. If some seminiferous tubules still exist, pipette them gently several times using a 10-ml glass pipette (plastic pipette is not suitable, see Note 2) without producing air bubbles and allow another 10~15 min for digestion by shaking at 34°C at 100 cycles/min.

24. The enzymatic digestion is checked again as step 22; if the seminiferous tubules are completely digested to single cells, 30 ml of fresh DMEM/F12 containing 10% FBS are added to stop the digestion; allow 10 min for sedimentation.

25. The supernatant containing germ cells and Sertoli cells is transferred to a new 50-ml conical tube and the sediment is discarded.

26. The supernatant is centrifuged at 112×g for 5 min and the cell pellet is suspended in 50 ml of fresh DMEM/F12.

27. After 10 min of sedimentation, the supernatant containing germ cells and Sertoli cells is transferred to a new 50-ml conical tube and the sediment is discarded.

28. The supernatant is centrifuged at 112×g for 5 min, and the cell pellet is suspended in fresh DMEM/F12 and filtered through a 40-μm nylon mesh to remove any cell aggregates.

29. The cell mixture is centrifuged at $112 \times g$ for 5 min and the cell pellet is suspended in fresh DMEM/F12 supplemented with 10% FBS.

3.3. Differential Plating to Obtain Human Male Germ Cells

For differential plating, the cell mixture containing male germ cells and Sertoli cells is cultured in DMEM/F12 supplemented with 10% FBS in a 15-cm diameter tissue culture dish precoated with 0.1% gelatin (see Note 6) for 3 h at 34°C. Sertoli cells attach to the culture plates, and the germ cells remain in suspension and are collected by centrifugation at $112 \times g$ for 5 min.

3.4. Isolation of Human GPR125- and GFRA1-Positive Spermatogonia by MACS

1. The germ cells are suspended in 1 ml of DMEM supplemented with 10% NU serum and 1 μg/μl DNase I (see Note 7).

2. A total of 25 μl of the antibody to GPR125 or GFRA1 is added to 1 ml cell solution and incubated at 4°C with rotation at a 50 cycles/min overnight.

3. After incubation with primary antibody, cell viability is checked with 0.4% trypan blue exclusion staining (see Note 8).

4. The cells are washed three times with BSA–EDTA–PBS buffer (ice-cold PBS containing 0.5% BSA and 2 mM EDTA) (see Note 9) for 5 min each time and are suspended in 80 μl of the BSA–EDTA–PBS buffer.

5. A total of 20 μl of goat anti-rabbit IgG magnetic microbeads is added to the cells and the cell mixture is incubated at 4°C with rotation at a 50 cycles/min for less than 20 min (see Note 10).

6. After incubation, 400 μl of BSA–EDTA–PBS buffer are added to the cells to get a final volume of 500 μl.

7. Around 5 μl of cell suspension is used to perform a viability test with 0.4% trypan blue exclusion staining.

8. The cells are loaded on the column that is equilibrated in the BSA–EDTA–PBS buffer.

9. The MACS preseparation filters (30 μm) are mounted on the column to filter out any aggregates before loading.

10. A total of 0.5 ml of cells is loaded on the column while the column is attached on the magnet.

11. The unlabelled cell fraction (i.e., GPR125- or GFRA1-negative cells) is first collected.

12. The column is washed three times with PBS–BSA–EDTA buffer (0.5 ml each time) and the unlabelled cell fraction (i.e., GPR125 or GFRA1-negative cells) is collected.

13. The column is removed from the magnet and GPR125- or GFRA1-positive spermatogonia are collected by adding 1 ml of BSA–EDTA–PBS buffer and pressing very gently with the plunger without making bubbles (see Note 11).

14. A second MACS is performed pursuant to the methods described as above so as to get a better purity of human GPR125- or GFRA1-positive spermatogonia.

3.5. Immunocyto-chemistry of Isolated Human GPR125- and GFRA1-Positive Spermatogonia

1. Human GPR125- or GFRA1-positive spermatogonia and GPR125- or GFRA1-negative male germ cells are washed with PBS and centrifuged at 112 g for 5 min and the cell pellet is suspended in PBS.

2. The cells are cytospun onto the microscope slides at 112 g for 5 min using a Cyto-Tek unit, and they are fixed with 4% para-formaldehyde for 30 min.

3. The cells are washed twice with PBS for 3 min at room temperature and blocked with 200 µl of normal 10% goat serum for 30 min at room temperature.

4. Primary antibody to GPR125, GFRA1, THY1, UCHL1, or ITGA6, or normal rabbit IgG, or normal mouse IgG, is diluted in 100 µl of PBS (dilution at 1:100–1:500); the diluted primary antibody is then added to the cells, and incubated for 1 h at 34°C or overnight at 4°C.

5. Replacement of the primary antibody with PBS serves as negative controls.

6. After incubation, the cells are washed with 200 µl of PBS.

7. After three washes in PBS, the cells are incubated with the secondary antibody, including Fluorescein (FITC)-conjugated goat anti-rabbit IgG, or Rhodamine-conjugated goat anti-rabbit or anti-mouse IgG.

8. DAPI is used to stain the nuclei and the cells are observed for epifluorescence using an Olympus Fluoview 500 Laser Scanning Microscope.

4. Notes

1. It is important to note that as soon as the blood supply to the testes is interrupted by "clamping" of the aorta, the Viaspan™ solution enter the testes.

2. It is essential to use glass pipettes to transfer the seminiferous tubules or male germ cells and Sertoli cells. Plastic pipettes are not suitable because the tubules or cells attach to the walls of plastic pipettes and significantly reduce the yield of male germ cells.

3. The time for the first digestion may vary and is dependent on the age of donors. Usually, 10 min are required for the first digestion, but it may take more time.

4. After the first digestion, extensive washes of the tubules (at least four times) with DMEM/F12 are required to remove the Leydig cells and other interstitial cells.

5. The time for the second digestion of the tubules to obtain single cells varies based on the age of the donors. Usually, 20 or 30 min are required, but it may take more time.

6. Tissue culture dishes are precoated with 0.1% gelatin, which facilitates Sertoli cell attachment to the dish within 3 h.

7. It is essential to adjust cell number to no more than ten million cells per ml of DMEM with 10% NU serum and DNase I.

8. Cell viability prior to MACS should be more than 95%. A higher percentage of dead cells can cause nonspecific binding to the separator column.

9. The BSA–EDTA–PBS buffer should be prefiltered and freshly prepared.

10. The cells cannot remain on ice with the microbeads for more than 20 min, since they are very sensitive at this stage and die if they are incubated for longer periods of time.

11. The cells can pass through the column without using the plunger. Only in the end, the plunger is placed on the top of the column and pressed gently and slowly for about 0.5 cm of the column until most of the liquid is out of the column. Bubbles in the column can kill the cells.

References

1. Nagano M, Brinster CJ, Orwig KE, Ryu BY, Avarbock MR, Brinster RL (2001) Transgenic mice produced by retroviral transduction of male germ-line stem cells. *Proc Natl Acad Sci USA.* **98**, 13090–13095.

2. Ryu BY, Orwig KE, Oatley JM, Avarbock MR, Brinster RL (2006) Effects of aging and niche microenvironment on spermatogonial stem cell self-renewal. *Stem Cells.* **24**, 1505–1511.

3. Brinster RL (2007) Male germline stem cells: from mice to men. *Science.* **316**, 404–405.

4. Kanatsu-Shinohara M, Ogonuki N, Iwano T, Lee J, Kazuki Y, Inoue K, Miki H, Takehashi M, Toyokuni S, Shinkai Y, Oshimura M, Ishino F, Ogura A, Shinohara T (2005) Genetic and epigenetic properties of mouse male germline stem cells during long-term culture. *Development.* **132**, 4155–4163.

5. He Z, Jiang J, Kokkinaki M, Golestaneh N, Hofmann MC, Dym M (2008) Gdnf upregulates c-Fos transcription via the Ras/Erk1/2 pathway to promote mouse spermatogonial stem cell proliferation. *Stem Cells.* **26**, 266–278.

6. Conrad S, Renninger M, Hennenlotter J, Wiesner T, Just L, Bonin M, Aicher W, Buhring H J, Mattheus U, Mack A, Wagner HJ, Minger S, Matzkies M, Reppel M, Hescheler J, Sievert KD, Stenzl A, Skutella T (2008) Generation of pluripotent stem cells from adult human testis. *Nature.* **456**, 344–349.

7. Guan K, Nayernia K, Maier LS, Wagner S, Dressel R, Lee JH, Nolte J, Wolf F, Li M, Engel W, Hasenfuss G (2006) Pluripotency of spermatogonial stem cells from adult mouse testis. *Nature.* **440**, 1199–1203.

8. Seandel M, James D, Shmelkov SV, Falciatori I, Kim J, Chavala S, Scherr DS, Zhang F, Torres R, Gale NW, Yancopoulos GD, Murphy A, Valenzuela DM, Hobbs RM, Pandolfi PP, Rafii S (2007) Generation of functional multipotent adult stem cells from GPR125+ germline progenitors. *Nature.* **449**, 346–350.

9. Kanatsu-Shinohara M, Inoue K, Lee J, Yoshimoto M, Ogonuki N, Miki H, Baba S, Kato T, Kazuki Y, Toyokuni S, Toyoshima M, Niwa O, Oshimura M, Heike T, Nakahata T, Ishino F, Ogura A, Shinohara T (2004)

Generation of pluripotent stem cells from neonatal mouse testis. *Cell.* **119**, 1001–1012.

10. Kanatsu-Shinohara M, Lee J, Inoue K, Ogonuki N, Miki H, Toyokuni S, Ikawa M, Nakamura T, Ogura A, Shinohara T (2008) Pluripotency of a single spermatogonial stem cell in mice. *Biol Reprod.* **78**, 681–687.

11. Ko K, Tapia N, Wu G, Kim JB, Bravo MJ, Sasse P, Glaser T, Ruau D, Han DW, Greber B, Hausdorfer K, Sebastiano V, Stehling M, Fleischmann BK, Brustle O, Zenke M, Scholer HR (2009) Induction of pluripotency in adult unipotent germline stem cells. *Cell Stem Cell.* **5**, 87–96.

12. Golestaneh N, Kokkinaki M, Pant D, Jiang J, Destefano D, Fernandez-Bueno C, Rone JD, Haddad BR, Gallicano GI, Dym M (2009) Pluripotent stem cells derived from adult human testes. *Stem Cells Dev.* **18**, 1115–1126.

13. Kossack N, Meneses J, Shefi S, Nguyen HN, Chavez S, Nicholas C, Gromoll J, Turek PJ, Reijo-Pera RA (2009) Isolation and characterization of pluripotent human spermatogonial stem cell-derived cells. *Stem Cells.* **27**, 138–149.

14. Mizrak SC, Chikhovskaya JV, Sadri-Ardekani H, Van Daalen S, Korver CM, Hovingh SE, Roepers-Gajadien HL, Raya A, Fluiter K, De Reijke TM, De La Rosette JJ, Knegt AC, Belmonte JC, Van Der Veen F, De Rooij DG, Repping S, Van Pelt AM (2010) Embryonic stem cell-like cells derived from adult human testis. *Hum Reprod.* **25**, 158–167.

15. Payne CJ, Braun RE (2008) Human adult testis-derived pluripotent stem cells: revealing plasticity from the germline. *Cell Stem Cell.* **3**, 471–472.

16. Geijsen N, Hochedlinger K (2009) gPS navigates germ cells to pluripotency. *Cell Stem Cell.* **5**, 3–4.

17. Naughton CK, Jain S, Strickland AM, Gupta A, Milbrandt J (2006) Glial cell-line derived neurotrophic factor-mediated RET signaling regulates spermatogonial stem cell fate. *Biology of Reproduction.* **74**, 314–321.

18. Hofmann MC, Braydich-Stolle L, Dym M (2005) Isolation of male germ-line stem cells; influence of GDNF. *Dev Biol.* **279**, 114–124.

19. Meng X, Lindahl M, Hyvonen ME, Parvinen M, De Rooij DG, Hess MW, Raatikainen-Ahokas A, Sainio K, Rauvala H, Lakso M, Pichel JG, Westphal H, Saarma M, Sariola H (2000) Regulation of cell fate decision of undifferentiated spermatogonia by GDNF. *Science.* **287**, 1489–1493.

20. He Z, Jiang J, Hofmann MC, Dym M (2007) Gfra1 silencing in mouse spermatogonial stem cells results in their differentiation via the inactivation of RET tyrosine kinase. *Biol Reprod.* **77**, 723–733.

21. Kubota H, Avarbock MR, Brinster RL (2003) Spermatogonial stem cells share some, but not all, phenotypic and functional characteristics with other stem cells. *Proc Natl Acad Sci USA.* **100**, 6487–6492.

22. Shinohara T, Avarbock MR, Brinster RL (1999) beta1- and alpha6-integrin are surface markers on mouse spermatogonial stem cells. *Proc Natl Acad Sci USA.* **96**, 5504–5509.

23. Luo J, Megee S, Rathi R, Dobrinski I (2006) Protein gene product 9.5 is a spermatogonia-specific marker in the pig testis: application to enrichment and culture of porcine spermatogonia. *Mol Reprod Dev.* **73**, 1531–1540.

24. He Z, Kokkinaki M, Jiang J, Dobrinski I, Dym M (2010) Isolation, characterization, and culture of human spermatogonia. *Biol Reprod.* **82**, 363–372.

25. Kubota H, Avarbock MR, Brinster RL (2004) Growth factors essential for self-renewal and expansion of mouse spermatogonial stem cells. *Proc Natl Acad Sci USA.* **101**, 16489–16494.

26. Kanatsu-Shinohara M, Miki H, Inoue K, Ogonuki N, Toyokuni S, Ogura A, Shinohara T (2005) Long-term culture of mouse male germline stem cells under serum-or feeder-free conditions. *Biol Reprod.* **72**, 985–991.

27. Clermont Y (1963) The cycle of the seminiferous epithelium in man. *Am J Anat.* **112**, 35–51.

28. Clermont Y (1966) Renewal of spermatogonia in man. *Am J Anat.* **118**, 509–524.

29. Clermont Y (1966) Spermatogenesis in man. A study of the spermatogonial population. *Fertil Steril.* **17**, 705–721.

30. Clermont Y (1972) Kinetics of spermatogenesis in mammals: seminiferous epithelium cycle and spermatogonial renewal. *Physiol Rev.* **52**, 198–236.

31. Dym M, Kokkinaki M, He Z (2009) Spermatogonial stem cells: mouse and human comparisons. *Birth Defects Res C Embryo Today.* **87**, 27–34.

32. Kokkinaki M, Lee TL, He Z, Jiang J, Golestaneh N, Hofmann MC, Chan WY, Dym M (2009) The molecular signature of spermatogonial stem/progenitor cells in the 6-day-old mouse testis. *Biol Reprod.* **80**, 707–717.

33. Kokkinaki M, Lee TL, He Z, Jiang J, Golestaneh N, Hofmann MC, Chan WY, Dym M (2010) Age affects gene expression in mouse spermatogonial stem/progenitor cells. *Reproduction.* **139**, 1011–1120.

Chapter 5

Isolation and Purification of Murine Male Germ Cells

Catherine Boucheron and Vanessa Baxendale

Abstract

Seminiferous tubules of the male testis contain somatic cells (Sertoli and Leydig cells) and germ cells at different stages of spermatogenesis (spermatogonia, spermatocytes and spermatids). Germ cells at different stages of differentiation migrate toward the central lumen via cell junctions formed by Sertoli cells. The protocol described herein consists of the dissection and decapsulation of the testes, disruption of the structure of the seminiferous tubules, and the breaking of cell junctions to release all the cells in suspension. Germ cells are then separated from Sertoli cells by overnight plating of the suspension on plastic to which the germ cells preferentially adhere. And finally, a BSA gradient allows a high-purity separation of the various types of germ cells according to size.

Key words: Germ cell, Testis, Spermatogonia, Gradient, Albumin

1. Introduction

Germ cells within the gonads are a unique entity, representing cells at various stages of the spermatogenic cycle. The testis provides unique microenvironments essential for the maintenance or developmental of each distinct stage of the male germ cell. Recently much attention has been devoted to the study of the more primitive germ cell, the spermatogonium. Many investigators have noted that the spermatogonium and its predecessor, the gonocyte, present a gene expression profile that is very similar to embryonic stem cells. For this reason and because they are relatively easy to obtain from adult animals, the testicular stem cell population has great potential for lateral comparison of genetic factors and mechanisms associated with early developmental processes.

Separation of the various cell populations in the testis requires knowledge of both tissue structure and cell morphology.

Wai-Yee Chan and Le Ann Blomberg (eds.), *Germline Development: Methods and Protocols*, Methods in Molecular Biology, vol. 825, DOI 10.1007/978-1-61779-436-0_5, © Springer Science+Business Media, LLC 2012

The method described herein is one of the most efficient protocols for the isolation of distinct cell populations from the testis that are highly homogeneous. The original method described by Bellve et al. in 1977 (1) and the modifications by Dym et al. in 1995 (2) have been combined to derive the protocol we use today. The modifications we have made allow the isolation germ cells from the testicular somatic cells at specific developmental times and their recovery at very high purity. As technology used for cellular and genetic analyses today is more sensitive, the homogeneity of the starting materials becomes a key factor for obtaining clear informative results.

2. Materials

2.1. Testis Isolation and Tissue Digestion

1. 1× RPMI 1640 (Invitrogen) supplemented with 1% Antibiotic–Antimycotic (Invitrogen) and 10% Fetal Bovine Serum (FBS; Invitrogen).

2. Collagenase Type 2 (Worthington Biochemicals): prepare a fresh stock of 2 mg/mL on the day of isolation using RPMI 1640. Store at 4°C until used.

3. Hyaluronidase (Sigma-Aldrich): prepare a stock of 10 mg/mL using RPMI 1640. Store aliquots of 1 mL at –20°C until used.

4. DNase I (Sigma-Aldrich): prepare a stock of 1 mg/mL using RPMI 1640. Store aliquots of 1 mL at –20°C until used.

5. Trypsin–EDTA 0.25% (Invitrogen). Store in aliquots of 6 mL at –20°C.

6. Sterile 0.22-μm nylon filters approximately 5 cm × 5 cm.

7. 150-mm culture plates (Corning).

8. CO_2 water jacketed incubator set to 34°C and 4% CO_2.

2.2. Bovine Serum Albumin Gradient

1. RPMI 1640 (Invitrogen).

2. Bovine serum albumin (BSA) fraction V lyophilized powder. 4 and 2% BSA solutions (w/v) are made using RPMI and filtered prior to use with a 0.45-μm syringe filter attached to a 50-mL syringe.

3. Gradient apparatus, see Fig. 1.

4. Siliconized 5-mL glass test tubes and rack.

2.3. Cell Identification and Counting

1. Hemocytometer.

2. Phase-contrast microscope – Zeiss Axiovert 200.

Fig. 1. Components of gradient apparatus. Items required are as follows: a gradient mixer capable of holding a minimum of 40 mL in each chamber, the 4% BSA solution is loaded in the chamber *on the left* while the 2% is added to the *right*; a 50-mL syringe; luer 3-way stop valve; siliconized 2.5-mm tubing; siliconized glass collection tubes.

3. Methods

3.1. Testis Isolation

The specific target cell one is interested in isolating will dictate the age mouse used with this procedure. For example, if one is interested in spermatogonia (both type A and type B spermatogonia), 6-day-old pups are more suitable than adult mice. By postnatal day 6, the first cycle of spermiogenesis has not yet begun in the mouse testis, therefore, germ cells at the later stages of differentiation are not present. If one is interested in a more differentiated stage of the spermatogonia, the adult mouse testis can be used for isolation. Once the testis is dissected from the animal, the capsule is mechanically removed using two pairs of iris forceps. The decapsulated testis is placed in ice-cold serum-free RPMI. The detailed procedure is described below.

1. Thirty male 6-day-old BALB/C mice (see Note 1) are anesthetized and sacrificed according to institutional animal care and use guidelines.

2. The testes are removed and pooled in 50 mL of ice-cold RPMI, in a 50-mL Falcon tube, on ice.

3. Once all testes are collected, each testicle in the pool is quickly decapsulated in a large culture plate containing fresh ice-cold RPMI, using two pairs of iris forceps (see Note 1).

4. The seminiferous cords are transferred into a new 50-mL tube containing 30 mL of fresh RPMI, on ice. After the decapsulated testis/seminiferous cords settle on the bottom of the tube, the supernatant is gently removed with a pipette.

3.2. Tissue Digestion

The progressive digestion of testicular tissue is the key to an efficient isolation of individual cells. There are three basic steps. In the first step, the seminiferous tubules are treated with a mixture of collagenase, hyaluronidase, and DNase I, to remove the interstitial cells, namely, myoid and Leydig cells (see Note 2). The second step of digestion with trypsin dissociates the seminiferous tubules and releases the Sertoli and germ cells. DNase I (see Note 3) is added to this step to reduce viscosity thus, cellular aggregation. Finally, the cell suspension is further purified, by panning; this exploits the fact that sertoli cells will adhere to the surface of the plastic culture dish while germ cells remain in suspension. At this point, tissue digestion and the primary separation of adherent and nonadherent cells is complete; the sample is ready for additional purification through a BSA gradient.

1. Incubate the seminiferous tubule pellet in a solution of 4.6 mL RPMI, 2.5 mL Collagenase (stock solution 2 mg/mL), 100 μL Hyaluronidase (stock solution 10 mg/mL), and 300 μL DNase I (see Note 4) (stock solution 1 mg/mL).

2. Gently swirl the sample by hand and place in a shaking water bath at 37°C for 30 min.

3. Remove sample from the shaker and allow the tubule tissue to settle to the bottom of the tube. Gently pipette off the supernatant containing the interstitial cells taking care not to remove the partially digested tubules.

4. Add 3 mL 0.25% Trypsin–EDTA (see Note 5) and 300 μL DNase I to the tubule tissue pellet, containing predominantly germ cells and Sertoli cells (see Note 6).

5. Swirl the sample by hand and place in a shaking water bath at 37°C for 15 min. Gently shake the tube every 5 min during this period.

6. Allow the tissue to settle to the bottom of the tube again. Transfer as much of the supernatant (cells in suspension) as can be without disturbing the larger chunks, into a fresh Falcon tube containing 20 mL of RPMI + 10% FBS on ice to inhibit proteolytic activity.

7. Add 3 mL 0.25% Trypsin–EDTA 3 and 300 μL DNase I to the remaining cell pellet, swirl by hand, and place in a shaking water bath for a 15-min incubation at 37°C. Shake the tube by hand every 5 min during this period to assist in dissociating the germ cells and the Sertoli cells.

8. Allow the tissues to settle by gravity and accumulate on the bottom of the tube.

9. Gently remove the supernatant and add to the previously prepared (step 8) 50-mL Falcon tube containing 20 mL of RPMI + 10% FBS and the first supernatant removed from the digestion. This pool should now contain predominantly germ and sertoli cells.

10. Filter the cell suspension through a sterile 0.22-μm nylon filter to remove any undigested fragments or cell clumps.

11. Centrifuge filtered cells at $800 \times g$ for 10 min at room temperature. Remove the supernatant and resuspend the cell pellet in 9 mL RPMI 1640.

12. Determine the cell concentration. Use 10 μL of the 9 mL suspension to count cells on a hemocytometer; you will obtain the total cell recovery value.

13. Prepare large cultures plates (150 mm) containing 30 mL RPMI + 5% FBS serum (15 mL RPMI added to 15 mL RPMI + 10% FBS). Divide the cell suspension between the plates, adding between 5×10^6 and 10×10^6 cells per plate (see Note 7). Incubate the cells overnight in a 34°C and 4% CO_2 water-jacketed CO_2 incubator, to allow adhesion of the Sertoli cells to the culture plate surface. This technique called panning is extremely efficient for the gross separation and purification of the germ cells.

3.3. BSA Gradient

Proper construction of the gradient is essential for effective cell separation. Utmost care should be taken to ensure that the gradient is both even and stable (see Note 8).

1. After the cells have been grossly separated into Sertoli cell (adhering to the culture plates) and germ cell (in suspension in culture medium) fractions, remove the supernatant, enriched with germ cells, and transfer to as many fresh 50-mL tubes as necessary (the number will depend of the number of culture plates used).

2. Centrifuge the supernatant for 10 min at $800 \times g$.

3. During centrifugation of the cells, prepare the BSA gradient (see Notes 9 and 10), using 30 mL of 4% and 30 mL of 2% BSA in RPMI 1640. Filter both BSA solutions before use with a 0.45-μm filter mounted on a 50-mL syringe.

4. Transfer the BSA solutions into the gradient maker apparatus. see Fig. 1. Set up of gradient apparatus. The gradient apparatus is connected to a 50-mL syringe in which the gradient is formed.

5. Open the gradient valve between the two chambers and allow the two BSA solutions to mix, and then open the release valve allowing the BSA to flow into the syringe to generate the gradient.

6. While the BSA gradient is stabilizing, remove the supernatant from the centrifuged of the cell suspension. Pool the pellets from several tubes together, and resuspend in RPMI 1640 to a total volume of 3 mL. Use 10 μL of the suspension to determine the cell number using a hemocytometer under light microscopy.

7. Gently overlay the top of the formed BSA gradient with the 3 mL cell suspension (see Note 11). Cover the top of the syringe containing cells/gradient with parafilm to minimize disturbance and changes in pH due to oxidation.

8. After 2.5 h at room temperature, collect the cells in 2 mL fractions eluting dropwise into siliconized 5-mL glass test tubes (see Note 12). On average, 25 fractions are obtained. Place the fractions on ice immediately after collection.

3.4. Cell Identification and Counting

1. Count the number of cells in each fraction under light microscopy using a hemocytometer using 10 μL of from each tube. Record the number of each cell type observed for the entire field. Determine the percentage of each type of cell (germ cells and Sertoli) based on size and morphology (Fig. 2) (see Notes 13 and 14).

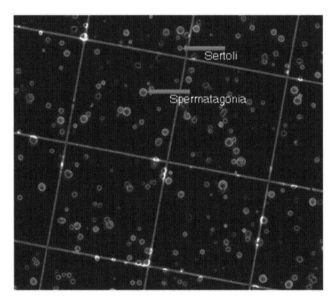

Fig. 2. Cell morphology. The morphologies of the cells present in the fractions are defined on the following criteria: Large cells presenting dense nuclei are spermatogonia while the smaller cells are either round spermatids or Sertoli cells. Depending upon the age of the mouse there may also be mid-spermiogenic stages which present as moderately large cells with more diffuse nuclei. This image is of a prefractionated sample from a 6-day-old mouse.

2. Pool the eluted fractions containing a high percentage of each type of germ cell and average the percent purity for that pool.

3. Centrifuge the sample at $800 \times g$ for 15 min. Remove the supernatant and store the pellet at −80°C until needed (see Note 15).

4. Confirm the identity of each cell type using specific cell markers.

4. Notes

1. We have also performed the experiment using both testicles from a single adult mouse, with minor modifications. The procedure described in this chapter uses 6-day-old mice in the isolation of spermatogonia as an example. Decapsulation of the testis is achieved by inserting the first set of forceps into the testis and opening then to generate a slit. The second pair is used to tease the tubules from the capsule gently.

2. Digestion of the seminiferous tubules must be done gently as excess agitation or enzymatic exposure will result in loss of germ cells due to early tubule degradation.

3. The use of DNase I is essential due to the high DNA content in the tissue; during the digestion, this keeps the sample from becoming gelatinous as the DNA becomes free in the solution.

4. The volume of DNAse I added in the digestion solutions for digestion of the adult mouse testicular tissue should be 500 μL for the first digestion and 450 μL for the subsequent two digestions. The rest of the protocol will be the same, however, it is important to note that the BSA gradient will contain all the different types of germ cells, and not only spermatogonia and Sertoli cells.

5. If the viscosity of the solution at this point is too high the cells will tend to stick together due to the increase of undigested DNA in the solution. If this happens, increase the volume of DNAse I by 200 μl added into the digestion mix.

6. Trypsin digestion is separated into two stages so as not to expose the germ cells to excessive trypsinization, which often results in unwanted cell lysis and impairs the Sertoli cells ability to adhere to the culture plate.

7. A concentration of cells either too low or too high will inhibit the ability of the Sertoli cells to adhere to the culture plates.

8. Once the gradient apparatus is set up, one should ensure the generation of a linear gradient using dye and a spectrophotometer.

9. We obtained better results with freshly prepared BSA solutions, compared to experiments performed with BSA solutions prepared the day before and stored at 4°C.

10. The gradient apparatus must be thoroughly washed after each use, first with ethanol, to avoid any contamination, and then three times with distilled water to remove any trace of ethanol, which could damage the cells.

11. Using a 1,000-μl pipette seems to be the best way to add the cell suspension without disturbing the gradient. The cells should form a definite layer on top of the column of BSA/RPMI.

12. Siliconized glass tubes are used to collect the eluted fractions to minimize cell loss due to cellular adhesion to the wall of the tube.

13. Under light microscopy spermatogonia appear large or medium, with a dense nucleus, while Sertoli cells appear much smaller and rounded. Other germ cell types such as pachytene spermatocytes and round spermatids can be identified by their distinct morphology (3).

14. In the purification of spermatogonia from 6-day-old mouse testis, the BSA gradient should roughly separate the cells into the following fraction ranges: Tubes 1–8 should contain large debris and cell clumps, tubes 9–15 should contain spermatogonia, and the last tubes should contain the smaller cells such as Sertoli cells and particulate matter.

15. If necessary, an additional resuspension of the cells in 1 mL fresh RPMI can be performed to count the number of cells in the final purified sample. 10 μL of this suspension can be used to record the number of spermatogonia using a hemocytometer. The sample is then centrifuged again for 10 min at $800 \times g$ and the pellet can be stored at −80°C until required for downstream analysis.

Acknowledgements

This work was supported by the Intramural Research Program of *Eunice Kennedy Shriver* National Institute of Child Health and Human Development, National Institutes of Health.

References

1. Bellvé AR, Millette CF, Bhatnagar YM, O'Brien DA. (1977) Dissociation of the mouse testis and characterization of isolated spermatogenic cells. J Histochem Cytochem 25:480–94.

2. Dym M, Jia MC, Dirami G, Price JM, Rabin SJ, Mocchetti I, Ravindranath N. (1995) Expression of c-kit receptor and its autophosphorylation in immature rat type A spermatogonia. Biol Reprod. 52:8–19.

3. Bellve AR, Cavicchia JC, Millette CF, O'Brien DA, Bhatnagar YM, Dym M. (1977) Spermatogenic cells of the prepubertal mouse. Isolation and morphological characterization. J. Cell Biol. 74:68–85.

Chapter 6

Preparation of Enriched Mouse Syncitia-Free Pachytene Spermatocyte Cell Suspensions

Yan Wang and Martine Culty

Abstract

Spermatogenesis comprises a complex succession of steps of mitosis, meiosis, and differentiation, starting with the commitment of diploid spermatogonial stem cells to differentiate and ending with the formation of haploid spermatozoa. Rodent models have been routinely used to study germ cell development and reproductive toxicology, since they present many similarities with their human counterpart while offering the advantage of recapitulating within a few weeks a process that normally takes years to occur. This article describes a method to isolate subpopulations of adult germ cells, more specifically pachytene spermatocytes, using two successive gradients, without using an elutriation centrifuge, a specialized device not available in many laboratories. Moreover, the method was designed to isolate enriched pachytene spermatocytes preparations devoid of contaminating syncitia, often formed in response to toxicants or environmental insults.

Key words: Testicular germ cell, Cell enrichment, BSA gradient, Percoll gradient, Pachytene spermatocytes

1. Introduction

Pachytene spermatocytes are formed in the prophase 1 of the first spermatogenic meiosis during which nonsister chromatids crossover (1). The isolation of pachytene spermatocytes can be complicated by the fact that a number of toxicants, chemical contraceptives, EDs, and diseases induce the formation of germ cell syncitia resulting from the fusion of different types of germ cells into giant multinucleated cells (2–5). Indeed, we found that these giant cells will elute in the same fractions as pachytene spermatocytes when using a STAPUT bovine serum albumin (BSA) gradient. Thus, the present method was developed to allow the isolation of enriched pachytene spermatocytes devoid of syncitia in sufficient numbers to study

Wai-Yee Chan and Le Ann Blomberg (eds.), *Germline Development: Methods and Protocols,* Methods in Molecular Biology, vol. 825, DOI 10.1007/978-1-61779-436-0_6, © Springer Science+Business Media, LLC 2012

the regulatory mechanisms involved in their survival and functions under different experimental conditions. One of the advantages of the present method is that it produces an excellent enrichment without using centrifugal elutriation, a process requiring a special centrifuge and rotors that have very limited use and therefore are not available in many laboratories. The method is a modified version of the original germ cell isolation method using a STAPUT gradient developed by Romrell et al. in 1976 (6). The main modification consists in the addition after the STAPUT gradient of an additional step of cell enrichment using a discontinuous Percoll gradient. Percoll is a nontoxic silica colloid used to form colloidal solutions of various densities which can be used to form either self-generated gradients by centrifugation or man-made discontinuous gradients. This gradient permits the separation of cells or particles according to their density by isopycnic centrifugation. We found that the addition of a Percoll gradient after the STAPUT gradient leads to highly enriched pachytene spermatocyte preparations despite their low abundance and the potential presence of syncitia in the total germ cell population. The cells are routinely isolated from the pooled testes of six to ten adult mice or rats. The principal of the method is to first isolate intact seminiferous cords from the interstitium by disruption of the interstitium by collagenase and hyaluronidase treatment, followed by further dissociation of the tubules with trypsin and collagenase. The different germ cell types are then separated by sedimentation velocity at unit gravity on a bovine serum albumin (BSA) gradient and the fractions containing pachytene spermatocytes are further enriched by separation on a Percoll density gradient.

2. Materials

2.1. Tissue Dissociation and Cell Isolation

1. All reagents have to be sterile or sterilized before use.
2. RPMI 1640 (RPMI) (Invitrogen) containing 100 U/ml penicillin and 100 mg/ml streptomycin (CellGro) supplemented with 0, 5 or 10% heat-inactivated Fetal Bovine Serum (FBS) (Invitrogen).
3. Collagenase: stock A: 2 mg/ml; stock B: 100 mg/ml, in RPMI + antibiotics; sterilize using 0.2-μm filters and freeze aliquots at –20°C (see Note 1).
4. Hyaluronidase: Type III testicular Hyaluronidase: 10 mg/ml RPMI + antibiotics; sterilize using 0.2-μm filters and freeze aliquots at –20°C.
5. DNAse I: stock A: 1 mg/ml; stock B: 10 mg/ml, both in RPMI + antibiotics; sterilize using 0.2-μm filters and freeze aliquots at –20°C.

6. Collagenase–DNAse solution: 4.7 ml RPMI+antibiotics, 2.5 ml collagenase, 0.8 ml DNAse I.

7. Trypsin–DNAse–Collagenase–Hyaluronidase solution: 7.65 ml commercial sterile solution of 0.25% Trypsin + 1 mM ethylene-diamine tetraacetic acid (EDTA), 0.1 ml DNAse I stock B, 0.25 ml collagenase stock B, 0.1 ml hyaluronidase.

2.2. BSA Gradient (Solutions Can Be Kept Overnight at 4°C)

1. 2% BSA solution: 5 g BSA fraction V (Roche) in 250 ml serum-free RPMI with antibiotics (filter solution using 0.2-μm filter).

2. 4% BSA solution: 10 g BSA fraction V (Roche) in 250 ml serum-free RPMI with antibiotics (filter solution using 0.2-μm filter).

3. Metricide (CSR): add 2.65 g Metricide powder to 500 ml solution.

4. Sigmacote solution: ready to use chlorinated organopolysiloxane in heptanes (Sigma). Used to silanize glassware: produces a neutral hydrophobic film on glass that will prevent adhesion of cells on surfaces.

2.3. Percoll Gradient

1. 10× saline: NaCl at 1.5 M (87.6 g/l).

2. Commercial Percoll stock: 23% (w/w) colloidal solution in water at a density of $1.130 + 0.005$ g/ml.

3. Percoll starting solution (PSS): 18 ml Percoll stock + 2 ml 10× saline.

4. 1× saline/Hepes (SH): 3 ml of 1.5 M NaCl + 26.85 ml H_2O + 0.2 ml Hepes 1 M, pH 7.4.

5. Components of the discontinuous gradient for a total gradient volume of 32 ml:

Prepare eight dilutions of Percoll solution as indicated in table below using different proportions of PSS and SH solutions (prepare ~5 ml of each dilution)

Tube #	1	2	3	4	5	6	7	8
Final density (g/ml)	1.08	1.07	1.06	1.055	1.050	1.04	1.035	1.03
PSS (ml)	3.5	3.0	2.5	2.5	2.0	1.5	1.25	1.25
SH (ml)	2	2.4	2.8	3.4	3.2	3.5	3.6	4.55
Volume of each dilution to use for 1 gradient (ml)	4	4	4	4	4	4	4	4

Use 4 ml of each dilution to form the successive layers of the discontinuous gradient, starting with the densest (#1) in bottom of tube to the least dense on top of the gradient, in a polycarbonate tube (plastic transparent enough to see through).

3. Methods (see Note 2)

3.1. Tissue Collection

1. Use six to ten adult male mice or rats per experiment. Animals are sacrificed by CO_2 inhalation according to the guide for the care and use of experimental animals from the US and Canadian Animal Care Agencies.

2. Prepare 50 ml RPMI + antibiotics kept in ice to collect and decapsulate testes.

3. Collect and keep testes in RPMI + antibiotics on ice for the duration of the dissection.

4. Clean testes from surrounding tissues and decapsulate.

5. Proceed to tissue dissociation in a culture hood under sterile conditions.

3.2. Tissue Dissociation

1. Collagenase–DNAse treatment: Add Collagenase–DNAse solution to the decapsulated testes.

2. Incubate testes for 20 min at 37°C in a shaking water bath (position tube in a way that allows efficient shaking of the tissues and maintenance of sterility). Decant.

3. Remove as much supernatant as possible without disturbing the tissue pellet. The supernatant contains interstitial cells, mainly Leydig cells, endothelial, and blood cells. It can be used to isolate Leydig cells by separating cell populations on a Percoll gradient.

4. The tissue pellet contains mainly seminiferous tubules. Wash the pellet twice by resuspending it in sterile PBS and letting the tubule pellet decant at each time, to remove trapped interstitial cells. Proceed for germ cell isolation.

5. Add the Trypsin–Collagenase–Hyaluronidase–DNAse I solution on the tissue pellet. Incubate 15 min at 37°C in the shaking water bath. Gently flush four to five times with pipette to perform mechanical dissociation (see Note 3).

6. Collect the supernatant into a tube containing 20 ml RPMI + antibiotics + 10% FBS to inhibit trypsin action. Filter cell mixture through 47 μm sterile nylon mesh to remove fragments. Keep cell suspension on ice.

7. Repeat steps 5 and 6 twice to by adding fresh enzyme solutions onto the tissue pellet, incubating for 5–10 min at 37°C in the shaking water bath, decanting, mixing supernatant with FBS and filtering.

8. Pool the filtered cell suspensions and centrifuge for 5 min at $800 \times g$ to pellet down the cells.

9. Remove the supernatant and add 5 ml RPMI + antibiotics to the resulting cell pellet.

10. Count cell numbers using a hemocytometer.

Number of cells in 16 squares $\times 10^4$ = number of cells / ml.

In general, one will get $50\text{--}200 \times 10^6$ germ cells per preparation. Adjust final volume to obtain 10^7 cells/ml.

3.3. Cell Separation on a BSA Gradient

1. The device is set up and the procedure performed in the cold room.

2. BSA (2–4%) gradient preparation: All equipment, gradient former, and tubing must be sterilized. Because this method uses a large volume glass STAPUT gradient former and sedimentation chamber, the glassware is siliconized with Sigmacote as indicated by manufacturer and autoclaved to avoid cell adhesion to the glass walls. Tubing is sterilized with a Metricide disinfectant solution (CSR) overnight. Rinse all equipment with sterile water and RPMI before use. The gradient former includes two compartments with a capacity of 300–500 ml each. A shallow gradient is formed in a large glass column that can contain 500 ml and is equipped with a small glass baffle at the bottom to avoid vortex and perturbations of the gradient.

3. Connect the output stopcock of the gradient former to a flexible tube ending on a three-way-stopcock attached to the gradient chamber in which the gradient will be formed. Another tube connects the third port of the syringe to a fraction collector.

4. In each of the two chambers of the gradient former, add 250 ml of 2% or 4% BSA solution, placing the 4% BSA in the back chamber. Add a magnetic bar in each chamber and place the gradient former on a magnetic stirring plate.

5. Open the connection between the front chamber and the sedimentation chamber to remove air bubbles in the tubing. Close the valve once a little amount of 2% BSA reaches the bottom of the sedimentation chamber.

6. Add the cell suspension at the bottom of the sedimentation chamber.

7. Open the connection between the back and front chambers of gradient former and the connection linking the gradient former to the sedimentation chamber. Open the valve to let the gradient slowly form, pushing the cell suspension toward the surface of the gradient as it is being made (see Note 4). Use a slow flow rate to avoid disruption of the gradient.

8. Let the cells sediment by gravity for 2 h to 2 1/2 h maximum.

9. Collect ~120 fractions of 4 ml each at a flow rate of 4 ml/min.

10. Most pachytene spermatocytes are found between fractions 10 and 35, but some come in the first fractions. Syncitia will elute in the first fractions too. Thus, one can usually collect and pool fractions 1–35 (see Note 5). However, because each gradient varies slightly, one should check fractions 36–46 with the hemocytometer for each gradient, to determine when the pachytene peak decreases and smaller cells start coming through and at which fraction to stop the pachytene pool collection.

11. Calculate the number of large (pachytene spermatocytes), medium (spermatogonia, leptotene and zygotene spermatocytes), and small (round spermatids) cells in the fractions using a hemocytometer. Large pachytene spermatocytes and small round spermatids are easily distinguished from other germ cell types. Estimate the relative proportion of each size group to determine the cutoff/border of each cell pool.

12. Pool the fractions containing the most pachytenes.

13. Most round spermatids elute between fractions 50 and 85. A good enrichment of round spermatids can be achieved by pooling fractions 55–85.

14. Fractions 36–49, between the peaks of pachytene spermatocytes and round spermatids, contain spermatogonia, leptotene and zygotene primary spermatocytes that do not form clear individual peaks.

3.4. Pachytene Spermatocyte Enrichment on Percoll Gradient

1. Centrifuge for 5 min at $800 \times g$ the pool of BSA gradient fractions containing the most pachytene spermatocyte (as determined in Subheadings 3.3.10 and 3.3.11).

2. Resuspend the cell pellet in 2 ml medium.

3. Apply the cell suspension gently and slowly on the top of the preformed Percoll gradient. Add water in a balance tube and equilibrate the tube weights on a balance.

4. Centrifuge tubes at $754 \times g$ for 20 min in a Sorvall refrigerated centrifuge using a fixed angle rotor at 4°C.

5. When finished, collect the fractions from the top of the gradient, separating clear zones from cloudy zones (if visible). If there is no clear visible band, collect 2 ml fractions, 17 fractions in total.

6. Check the cell sizes and numbers for each Percoll fraction with a hemocytometer. Pool adjacent fractions that contain similar cells (by size and morphology on microscope) (Fig. 1).

7. Pachytene spermatocytes are found in fractions 10–14 (peak at 11) of the gradient. Large syncitia are found in fractions 1–6. Around 800,000–1 million cells can be obtained per experiment.

8. Each fraction or pool is diluted (1:5) with 5 volumes of 1× saline to dilute the Percoll. Cells are collected by centrifugation

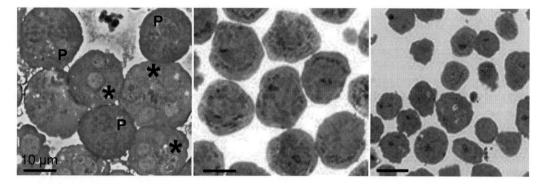

Fig. 1. Morphological appearance of isolated germ cells. Fractions from BSA gradient (*left and right panels*) and Percoll gradient (*middle panel*) were embedded in resin, sectioned, and observed under a light microscope. *Left panel*: fraction from BSA gradient showing the presence of multinucleated germ cells (*) and pachytene spermatocytes (P). *Middle panel*: fraction from Percoll gradient showing enrichment in pachytene spermatocytes. *Right panel*: as a comparison, other types of smaller sizes germ cells. All scales are 10 μm.

at $200 \times g$ for 20 min. Cell pellets are frozen at –80°C for mRNA or protein analysis.

4. Notes

1. The dilutions of collagenase, hyaluronidase, and DNAse I might need to be adjusted depending of the specific activity of the enzyme lots purchased. For example, an insufficient amount/activity of DNAse can lead to the tubule pellets becoming viscous, a condition that hampers cell isolation and leads to poor yields.

2. Before starting, be aware that this is a lengthy procedure that takes around 12 h without interruption from start to end. In contrast with the isolation of neonatal germ cells, there is no overnight adhesion phase in the adult cells isolation procedure, since there are too few Sertoli and myoid cells to justify this step.

3. Light mechanical dissociation at the end of trypsinization will help. However, do not flush the tissue too much with the pipet or viscous tissue aggregates will form leading to poor cell recovery after trypsinization.

4. Alternatively, some STAPUT devices include an additional column between the gradient former and sedimentation chamber where the cell suspension is placed that will allow loading the cell suspension on the top of the gradient as it is being made.

5. In the initial experiment, it might be helpful to observe every two to three fractions with the hemocytometer and to count

cells to get a good idea of the size differences and to estimate the proportion of each size group. This will help appreciating the size difference between the main cell types. Moreover, observing the profile of each peak through the observation of many fractions will help better assessing where each cell pool should start and end.

Acknowledgements

This work was supported in part by a Centennial award from the Royal Victoria Hospital foundation, Montreal, and by a grant (ES 10366) from the National Institutes of Environmental Health Sciences, National Institutes of Health to M.C. The Research Institute of MUHC is supported in part by a Center grant from Le Fonds de la Recherche en Santé du Quebec. We thank Corinne Rhodes for her outstanding technical assistance.

References

1. Hermo l, Pelletier RM, Cyr DG, Smith CE (2010) Surfing the Wave, Cycle, Life History, and Genes/Proteins Expressed by Testicular Germ Cells. Part 1: Background to Spermatogenesis, Spermatogonia, and Spermatocytes. Microscopy Research and Technique; 73: 243–278

2. Faridha A, Faisal K, Akbarsha MA (2007) Aflatoxin treatment brings about generation of multinucleate giant spermatids (symplasts) through opening of cytoplasmic bridges: Light and transmission electron microscopic study in Swiss mouse. Reproductive Toxicology; 24:403–408

3. Hild SA, Reel JR, Dykstra MJ, Mann PC, Marshall GR (2007) Acute adverse effects of the indenopyridine CDB-4022 on the ultrastructure of sertoli cells, spermatocytes, and spermatids in rat testes: comparison to the known sertoli cell toxicant Di-n-pentylphthalate (DPP). J Androl; 28:621–629

4. Singh A, Singh SK (2009) Evaluation of antifertility potential of Brahmi in male mouse. Contraception 79:71–79

5. Pelletier RM, Yoon SR, Akpovi CD, Silvas E, Vitale ML (2009) Defects in the regulatory clearance mechanisms favor the breakdown of self-tolerance during spontaneous autoimmune orchitis. Am J Physiol Regul Integr Comp Physiol; 296:R743–R762

6. Romrell, L. J., A. R. Bellvé, and D. W. Fawcett (1976) Separation of mouse spermatogenic cells by sedimentation velocity. A morphological characterization. Dev Biol; 49:119–131

Chapter 7

Revealing the Transcriptome Landscape of Mouse Spermatogonial Cells by Tiling Microarray

Tin-Lap Lee, Owen M. Rennert, and Wai-Yee Chan

Abstract

In the past decade, the advent of microarray technologies has allowed functional genomic studies of male germ cell development, resulting in the identification of genes governing various processes. A major limitation with conventional gene expression microarray is that results obtained are biased due to gene probe design. The gene probes for expression microarrays are usually represented by a small number of probes located at the 3′ end of a transcript. Tiling microarrays eliminate such issue by interrogating the genome in an unbiased fashion through probes tiled across the entire genome. These arrays provide higher genomic resolution and allow the identification of novel transcripts. To reveal the complexity of the genomic landscape of developing male germ cells, we applied tiling microarray to evaluate the transcriptome in spermatogonial cells. Over 50% of all known mouse genes are expressed during testicular development. More than 47% of the transcripts are uncharacterized. The results suggested that the transcription machinery in spermaotogonial cells is more complex than previously envisioned.

Key words: Male germ cells, Development, Transcriptome, Tiling microarray, Expression profiling

1. Introduction

The advent of microarray technologies provides solutions to study the gene expression profile in germ cell development. A major observation from these microarray studies is that the genome is actively transcribed during germ cell development (1–3). Importantly, many differentially expressed transcripts are unknown or uncharacterized (4–6). In addition, the germ cell transcriptome exhibits dynamic changes in conjunction with specific developmental regulation. Male germ cell development happens in three phases. The first phase, from day 0 to postnatal day 8, is mitosis when spermatogonial proliferation predominates. The second phase occurs at the initiation

Wai-Yee Chan and Le Ann Blomberg (eds.), *Germline Development: Methods and Protocols*, Methods in Molecular Biology, vol. 825, DOI 10.1007/978-1-61779-436-0_7, © Springer Science+Business Media, LLC 2012

of meiosis, on day 14, during early pachytene spermatocyte development. This is followed by entry into spermiogenesis on day 20 when round spermatids first appear (7). The dynamic and specific nature of the germ cell transcriptome has been shown to be associated with specific development and regulatory programs. Ontology analysis of the germ cell transcriptome reveals different biological processes distinctively associated with mitotic, meiotic, and postmeiotic male germ cells, respectively (2, 7–9).

Although microarrays prove to be an invaluable tool in the functional genomics study of germ cell development, it has limitations. Traditional microarrays represent only a select set of transcripts at the time of fabrication, and only a small number of probe sets determine the expression level of a given transcript. Transcripts with alternative splicing isoforms may be misinterpreted, if the probes are all located in the common region shared by the variants. The probes in traditional microarray platforms are designed based on sequences located at the 3′ end region of a transcript. Therefore, they may not provide a complete and accurate picture of transcript expression data. This presents a challenge in downstream validation experiments.

Tiling microarrays address these issues by designing probes throughout the entire genome or contigs of the genome in an unbiased fashion with a higher resolution (10–14). The whole genome coverage feature allows a wide range of genomic applications, including whole genome transcriptome analysis (14), detection of transcription factor-binding sites (13), chromatin modifications, and epigenetic modifications, such as DNA or histone methylation (15). In this chapter, we apply Affymetrix mouse tiling 1.1R microarray sets (GeneChips) to examine the transcriptional landscape in murine spermatogonial cells.

The procedure from capturing polyadenylated transcripts to microarray scanning is illustrated in Fig. 1. Basically, the mRNA population in the total RNA is first enriched by using the RiboMinus kit. The enriched mRNA population is then converted to cDNA. The cDNA product serves as a template for subsequent in vitro transcription to cRNA. Contrary to conventional 3′ in vitro transcription (IVT) GeneChip microarrays, the cRNA is amplified for another round to cDNA before fragmentation and labeling. Once the fragmented products are labeled with biotin, they are mixed with eukaryotic RNA and oligo B2 controls for array hybridization.

Overall, we found that more than 45% of transcripts were not annotated. Current annotation accounts for only about 30% of the data set. Majority of data set is in the form of expressed sequence tags (ESTs) or noncoding RNAs.

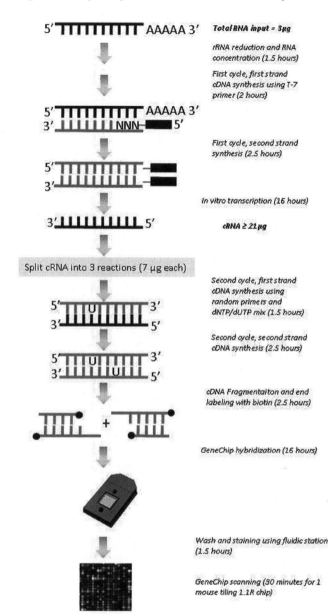

Fig. 1. Overview of tiling microarray protocol for Affymetrix Mouse 1.1R GeneChip. The mRNA population from input total RNA is first enriched by rRNA reduction. Isolated mRNA is then reverse transcribed to cDNA and amplified by in vitro transcription into cRNA. The cRNA population is split into three reactions for another round of cDNA synthesis. The cDNA is subsequently fragmented and labeled, and ready for hybridization with GeneChip. Hybridization mix can be reused for up to three times. Therefore, the protocol is sufficient for nine GeneChip hybridization. The protocol takes three days to complete.

2. Materials

2.1. Spermatogonial Cell Samples

1. Type A spermatogonia were isolated from 6-day-old Balb/c mice. Since 6-day-old mice testis do not have spermatocytes or spermatids, type A spermatogonia are typically >99% pure after separation from contaminating Sertoli cells by differential plating.

2. For transcriptome tiling array application in triplicates, at least 15 μg of total RNA is required. The minimum concentration is 0.24 μg/μl.

3. RNA quality is critical to the overall success of experiment. Minimize the number of freeze and thaw cycle if possible and make sure the experiment is performed in RNAase-free environment.

2.2. Tiling Microarrays

1. Affymetrix GeneChip® Mouse Tiling 1.1R Array Set designed for whole transcriptome analysis.

2. The chips are based on Affymetrix 49 format. Each chip contains more than 3 million probe pairs (match and mismatch) to cover whole mouse genome at an average of 35 base-pair resolution. The average gap between probes is 10 base pair. All repetitive elements derived from RepeatMasker were removed.

3. Sequences used in probe set design were based on NCBI mouse genome assembly (Build33). The results could be updated to other build version using genome coordinate conversion tools (see methods).

2.3. rRNA Reduction

1. RiboMinus Human/Mouse Transcriptome Isolation Kit. Resuspend the RioMinus Magnetic Beads completely before use.

2. Magna-Sep Magnetic Particle Separator.

3. 5 M betaine.

2.4. Synthesis, Amplification, Labeling, Cleanup, and Fragmentation of cDNA and cRNA

1. GeneChip WT Amplified Double-Stranded cDNA Synthesis Kit containing:

   2.5 μg/μl T7 primers, 5× first-strand buffer, 0.1 M DTT, 10 mM dNTP, RNase inhibitor, SuperScript II, 1 M MgCl$_2$, DNA polymerase I, RNase H, 3 μg/μl random primers, 10 mM dNTP and dUTP, and RNase-free water.

2. GeneChip WT cDNA Amplification Kit containing:

   10× IVT buffer, IVT NTP mix, IVT enzyme mix, and IVT control.

3. GeneChip WT Double-Strand DNA terminal Labeling Kit containing:

   5× cDNA fragmentation buffer, 10 U/μl UDG, 100 U/μl APE1, 5× TdT buffer, 30 U/μl TdT, 5 mM GeneChip DNA-labeling reagent, and RNase-free water.

4. GeneChip Sample Cleanup Module containing:

cDNA cleanup spin columns, cDNA-binding buffer, 6 ml concentrate cDNA wash buffer, cDNA elution buffer, IVT cRNA cleanup spin columns, IVT cRNA-binding buffer, 5 ml concentrate IVT cRNA wash buffer, and 5× fragmentation buffer.

2.5. Gel-Shift Assay

1. RNA 6000 Nano LabChip Reagents and Supplies (Agilent).
2. 4–20% TBE Gel, 1 mm, 12 well (Invitrogen).

2.6. GeneChip Hybridization, Wash, and Staining

1. GeneChip Hybridization and Wash kit: 2× prehybridization mix, hybridization mix, and DMSO; store at room temperature.
2. Stain cocktail 1, light sensitive, stored in dark bottle.
3. Stain cocktail 2.
4. Array holding buffer: Wash buffer A, stored at room temperature to prevent formation of salt particles, wash buffer B, light sensitive, stored in dark bottle.
5. 3 nM control oligonucleotide B2.
6. 20× eukaryotic hybridization controls (bioB, bioC, BioD, cre): Frozen stocks heated to 65°C for 5 min to resuspend the cRNA completely before use.

2.7. Instrumentation

1. GeneChip Scanner 3000 7 G: Previous versions of 3000 series scanner can be upgraded to 7 G through Affymetrix scanner upgrade program.
2. GeneChip Hybridization Oven 640.
3. GeneChip AutoLoader (optional).
4. GeneChip Fluidics Station 400 or 450 series. This protocol in this chapter is based on Fluidics 450 station.
5. NanoDrop ND-1000 (Ambion).
6. Bioanalyzer 2100 (Agilent).

2.8. Software

1. Affymetrix GeneChip Command Console Software (AGCC) or GeneChip Operating Software (GCOS) version 1.3 or higher.
2. Affymetrix Tiling Analysis Software (TAS).
3. Integrated Genome Browser from Web start (http://www. affymetrix.com/partners_programs/programs/developer/ tools/download_igb.affx) or through local installation.
4. The software are available for free for Affymetrix users, downloadable at http://www.affymetrix.com.

3. Methods

3.1. Enriching mRNA Fraction from Total RNA Using RiboMinus Kit

1. Prepare RiboMinus buffer with betaine by mixing 120 μl of betaine (5 M) and 280 μl of RiboMinus hybridization buffer.

2. Prepare RiboMinus probe hybridization mix in a 0.2-ml strip tube according to the table below and incubate at 70°C for 5 min in a thermal cycler.

	Concentration of total RNA input	
	≥0.4 μg/μl	≤0.4 μg/μl and ≥0.24 μg/μl
Total RNA	3 μg	3 μg
RiboMinus Probe	2.4 μl	2.4 μl
Hybridizaation buffer with betaine	60 μl	90 μl
RNase-free water	up to 70 μl	up to 105 μl
Total volume	70 μl	105 μl

3. Meanwhile, preheat a 37°C heat block and a 50°C heat block for mRNA capture using magnetic beads.

4. Resuspend the RiboMinus magnetic beads by pipetting several times until no sediment is observed at the bottom. Transfer 150 μl of beads to a 1.5-ml nonstick RNase-free tube (see Note 1).

5. Put the tube on a magnetic stand. Aspirate and discard the clear supernatant after 1 min.

6. Wash the magnetic bead pellet with 150 μl RNase-free water by resuspending the pellet. Spin briefly and put the tube back to the magnetic stand for 1 min. Repeat the step one more time.

7. Wash with 150 μl of hybridization buffer with betaine. Aspirate and discard the supernatant.

8. Depending on the concentration of the input RNA, add 90 (input ≥0.4 μg/μl) or 60 (less than 0.4 μg/μl) μl of hybridization buffer with betaine to resuspend the magnetic bead pellet.

9. Incubate the mix in the 37°C heat block for 5 min, resuspend the mix, and incubate for 5 more minutes. Place the tube in the magnetic stand for 1 min.

10. Transfer the clear supernatant containing the purified RNA to a 1.5-ml nonstick RNase-free tube and leave on ice.

11. Wash the magnetic beads with 50 μl of hybridization buffer with betaine, and incubate in the 50°C heat block for 5 min. Place the tube back to magnetic stand for 1 min.

12. Aspirate the clear supernatant and transfer to the tube containing the purified RNA in step 10.

3.2. RNA Cleanup

1. Add 735 µl of cRNA-binding buffer from the GeneChip Sample Cleanup Module to each sample and vortex for 3 s. Then, add 525 µl of 100% ethanol to each sample. Mix by flicking.

2. Apply the sample to an IVT cRNA column in a 2-ml collection tube.

3. Centrifuge for 15 s at $8,000 \times g$. Discard the flow through.

4. Using the same column, repeat steps 2 and 3 for the remaining samples.

5. Discard the collection tube and combine the column in a new 2-ml collection tube. Add 500 µl of cRNA wash buffer and centrifuge at $8,000 \times g$ for 15 s. Discard the flow through.

6. Add 500 µl of 80% ethanol and centrifuge at $8,000 \times g$ for 15 s. Discard the flow through.

7. Open the column cap and centrifuge at maximum speed for 5 min.

8. Transfer the column to a new 1.5-ml collection tube. Add 11 µl of RNase-free water directly to the center of the membrane. Centrifuge at maximum speed for 1 min.

9. The eluted RNA should be around 9.8 µl. Analyze the quality of purified RNA using a bioanalyzer (optional, see Notes 2 and 3).

3.3. First-Cycle, Double-Stranded cDNA Synthesis

1. Prepare the RNA/T7 primer mix by combining 4 µl of RNA and 1 µl of 500 ng/µl T7 primer (see Note 4).

2. Incubate at 70°C for 5 min. Place the sample on ice.

3. Prepare the first-cycle, first-strand master mix as follows:

Component	Volume in 1 reaction
5× first-strand buffer	2 µl
0.1 M DTT	1 µl
10 mM dNTP mix	0.5 µl
RNase inhibitor	0.5 µl
SuperScript II	1 µl
Total volume	5 µl

4. Add 5 µl of master mix to each sample to make a total volume of 10 µl.

5. Incubate the samples in a thermal cycler as follows: 25°C for 10 min; 42°C for 1 h; 70°C for 10 min; and 4°C for 10 min.

6. Prepare the second-cycle, second-strand master mix as follows:

Component	Volume in 1 reaction
17.5 mM $MgCl_2$	4 µl
10 mM dNTP mix	0.4 µl

Component	Volume in 1 reaction
DNA polymerase I	0.6 μl
RNase H	0.2 μl
RNase-free water	4.8 μl
Total volume	10 μl

7. Combine the first-cycle, first-strand mix from step 5 to the mix from step 6 to make a total volume of 20 μl.

8. Incubate the samples in a thermal cycler as follows: 16°C for 2 h without heated lid; 75°C for 10 min with heated lid; and 4°C for 10 min.

3.4. In Vitro Transcription Amplification

1. Prepare the IVT master mix as follows:

Component	Volume in 1 reaction
10× IVT buffer	5 μl
IVT NTP mix	20 μl
IVT enzyme mix	5 μl
Total volume	30 μl

2. Add 30 μl of IVT master mix to each sample to make a total volume of 50 μl. Mix and briefly spin down the mix.

3. Incubate the samples in a thermal cycler as follows: 37°C for 16 h and 4°C at hold position.

4. On day 2, clean up the cRNA from step 3 using the cRNA cleanup spin columns from the cleanup module.

5. Add 50 μl of RNase-free water to each sample to make a total volume of 100 μl.

6. Add 350 μl of cRNA-binding buffer to each sample and vortex for 3 s.

7. Add 250 μl of 100% ethanol to each sample and flick mix.

8. Transfer the sample to the IVT cRNA column in a 2-ml collection tube.

9. Centrifuge at $8,000 \times g$ for 15 s. Discard the flow through.

10. Discard the collection tube and combine the column to a new 2-ml collection tube. Add 500 μl of cRNA wash buffer and centrifuge at $8,000 \times g$ for 15 s. Discard the flow through.

11. Add 500 μl of 80% ethanol and centrifuge at $8,000 \times g$ for 15 s. Discard the flow through.

12. Open the column cap and centrifuge at maximum speed for 5 min.

13. Transfer the IVT cRNA column to a new 1.5-ml collection tube. Add 12 µl of RNase-free water directly to the center of the membrane. Centrifuge at maximum speed for 1 min.

14. Repeat step 13. The eluted cRNA is about 21 µl. Analyze the quality of purified RNA by measuring the readings at 260, 280, and 320 nm using spectrophotometer. The concentration of cRNA is calculated by this formula (see Note 5):

$$[A_{260} - A_{320}] \times 0.04 \times \text{dilution factor.}$$

3.5. Second-Cycle, Double-Strand cDNA Synthesis

1. Divide each sample into three separate reactions to be used for hybridization with three tiling microarrays by preparing the second-cycle cRNA/random primer mix as follows:

Component	Volume in 1 reaction
7 µg cRNA	Variable
3 µg/µl random primers	1 µl
RNase-free water	Up to 8 µl
Total volume	8 µl

2. Flick, mix, and spin down the tubes.

3. Incubate the samples in a thermal cycler as follows: 70°C for 5 min; 25°C for 5 min; and 4°C for 10 min.

4. Prepare second-cycle, first-strand cDNA synthesis master mix as follows:

Component	Volume in 1 reaction
5× first-strand buffer	4 µl
0.1 M DTT	2 µl
10 mM dNTP and dUTP	1.25 µl
SuperScript II	4.75 µl
Total volume	12 µl

5. Combine the first-cycle, first-strand cDNA synthesis master mix with cRNA/random primer mix from step 1 to make a total volume of 20 µl.

6. Incubate the samples in a thermal cycler as follows: 25°C for 10 min; 42°C for 90 min; 70°C for 10 min; and 4°C for 10 min.

7. Prepare second-cycle, second-strand cDNA synthesis master mix as follows:

Component	Volume in 1 reaction
17.5 mM MgCl$_2$	8 μl
10 mM dNTP and dUTP	1 μl
DNA polymerase I	1.2 μl
RNase H	0.5 μl
RNase-free water	9.3 μl
Total volume	20 μl

8. Combine the second-cycle, second-strand cDNA synthesis master mix with second-cycle, first-strand cDNA reaction from step 6 to make a total volume of 40 μl.

9. Incubate the samples in a thermal cycler as follows: 16°C for 2 h without heated lid; 75°C for 10 min with heated lid; and 4°C for 10 min.

10. Clean up the cDNA using cDNA cleanup spin columns from the GeneChip Sample Cleanup Module.

11. Add 60 μl of RNase-free water and 370 μl of cDNA-binding buffer to each cDNA sample. Vortex for 3 s.

12. Transfer the mix to the cDNA-binding column with 2-ml collection tube and centrifuge at 8,000×g for 15 s. Discard the flow through.

13. Add 500 μl of 80% ethanol and centrifuge at 8,000×g for 15 s. Discard the flow through.

14. Discard the collection tube and combine the cDNA column to a new 2-ml collection tube. Add 750 μl of cDNA wash buffer and centrifuge at 8,000×g for 15 s. Discard the flow through.

15. Open the column cap and centrifuge at maximum speed for 5 min.

16. Transfer the cDNA column to a new 1.5-ml collection tube. Add 15 μl of RNase-free water directly to the center of the membrane. Centrifuge at maximum speed for 1 min.

17. Repeat step 16. The eluted cDNA is about 28 μl. Analyze the quality of purified cDNA by measuring the readings at 260, 280, and 320 nm using spectrophotometer. The concentration of cDNA is calculated by this formula: $(A_{260}-A_{320}) \times 0.05 \times$ dilution factor.

18. Each sample tube should contain at least 7.5 μg of cDNA. Pooling the cDNA sample is recommended to minimize variability across reactions.

3.6. Fragmentation and Labeling

1. Fragment the cDNA by preparing the fragmentation mix as follows:

Component	Volume in 1 reaction
cDNA	7.5 μg
10× fragmentation buffer	4.8 μl
10 U/μl UDG	1.5 μl
100 U/μl APE1	2.25 μl
RNase-free water	Up to 48 μl
Total volume	48 μl

2. Incubate the samples in a thermal cycler as follows: 37°C for 1 h; 93°C for 2 min; and 4°C for 10 min.

3. Transfer 45 μl of fragmented sample to a new tube and use the remaining for fragmentation analysis using RNA 6000 Nano LabChip.

4. Label the fragmented cDNA by preparing the labeling mix as follows:

Component	Volume in 1 reaction
Fragmented cDNA	45 μl
5× TdT buffer	4.8 μl
30 U/μl TdT	2 μl
5 mM DNA labeling reagent	1 μl
Total volume	60 μl

5. Incubate the samples in a thermal cycler as follows: 37°C for 1 h; 70°C for 10 min; and 4°C for 10 min.

6. Take 4 μl for gel-shift analysis (optional, see Note 6).

3.7. Hybridization and Scanning

1. Prepare hybridization cocktail as follows:

Component	Volume in 1 reaction
Labled target	60 μl
Control oilgonucleotide B2	4.17 μl
2× hybridization mix	125 μl
DMSO	17.5 μl
RNase-free water	Up to 250 μl
Total volume	250 μl

2. Heat the cocktail at 99°C for 5 min. Cool to 45°C for 5 min and centrifuge at maximum speed for 1 min.

3. Inject 200 μl of cocktail into the tiling array chip (see Note 7).

4. Place the chips in the hybridization oven at 45°C. Set to rotate at 60 rpm (0.45× g). Incubate for 16 h.

5. Remove and save the hybridization cocktail after incubation. Refill the chip with 250 μl of wash A buffer.

6. Set up the Fluidic Station 450 and apply FS450_0001 protocol. Place 600 μl of stain cocktail 1 in sample holder 1. Place 600 μl of stain cocktail 2 in sample holder 2 and 800 μl of array-holding buffer in sample holder 3 (see Note 8).

7. Load the chip to scanner when the fluidic procedure is finished (see Note 9).

3.8. Data Analysis

The data analysis workflow in this section is based on Tiling Analysis (TAS) software provided by Affymetrix. It provides basic analysis functions, including raw data processing to generate normalized intensity values and p-values for the probes, genomic interval computation, and visualization for evaluating the quality of array data. The analysis can be further processed in downstream software pipeline, such as IGB and UCSC genome browser. Here, we show how to process and analyze the scanned raw intensity data in CEL format (see Notes 10 and 11).

3.8.1. Define Analysis Group

1. Create a folder for tiling array analysis (e.g., C:\TMA-workflow) and put the CEL intensity data file in this folder. Then, define the sample set in TAS software through generation of a Tiling Analysis Group (TAG) file (see Note 12). A total of 14 TAG files (Chip A to Chip N) are generated to represent a complete transcriptome.

2. Identify the CEL replicates from the same chip. To associate the probe set intensity to corresponding genomic position, the user needs to supply genomic location (bpmap) file.

3. Select "Edit→default" and press "paths" tab in TAS to locate the bpmap file. Make sure that the file is properly uncompressed.

3.8.2. Data Normalization

1. Normalization is required for replicate data. It helps mitigating signal intensity differences caused by variations in biological samples and target preparation process.

2. This method is applied in our transcriptome data analysis (see Note 13).

3. Select "Edit→default" and press "paths" tab in TAS to locate the directories for the data and library files.

3.8.3. Intensity Analysis

1. Once data are normalized, the intensity values signal and p-values for each probe are calculated. Calculate the window size by 2× bandwidth +1 (see Note 14).

2. For transcriptome analysis, we use a window size of 151 bp (see Note 15). Select "Edit→default" and press "Probe Analysis" and put 75 in the bandwidth window.

3. Apply one sample analysis to calculate the p-values based on signals from PM and MM probe sets (see Note 16). The p-values are presented in $-10\log_{10}$ (p-value), which means a p-value of 0.1 is equal to 10 and a p-value of 0.01 is equal to 20. This is done by selecting "Edit→default" in TAS, pressing "Probe Analysis," and selecting "One side Upper" under "Test Type" field and "PM/MM" under "intensity" field.

4. To generate the intensity output files consisting of signal intensity and p-values, select "Edit→default" and press "Export" tab in TAS. Under "Results section," select "Save signal and p-values," the files are saved separately. Save the files in .bar file format by selecting "Save to BAR format" in "Signal/p-values Output File" section (see Note 17).

5. Proceed to "Scale" in the "Default Properties" window to select scale of the data. Select "Log2" under "Signal Scale" and "$-10\log_{10}$" under "p-value Scale," and click "OK" to finish.

6. To run intensity analysis, select "Analyze→Intensities" in the menu of TAS. Open the TAG file created in define analysis group step to start analysis.

3.8.4. Assessing Intensity Data

1. To view the quality of transcriptome data, start IGB browser (see Note 18). Depending on the scale of analysis, select the memory requirement of IGB according to the IGB manual.

2. Select the genome and build version corresponding to the bpmap file applied in TAS by accessing the "Data Access" tab in the lower window of IGB; the Refseq gene annotation and associated genome coordinates are retrieved in the upper window.

3. Choose the chromosome corresponding to the intensity or p-value data file by highlighting the sequence under "Current genome." The table below shows the chromosome information contained in the mouse tiling 1.1R microarray.

4. To navigate the genome, use the scroll bar on the bottom of the viewer to move in either direction along the genome. Use the arrows on both sides for fine movements. To zoom in or out at a specific position, slide the scroll bar on the top window of IGB.

5. The location of a gene can be retrieved by entering genomic coordinate or gene symbol under the "Pattern Search" tab in the lower window (see Note 19).

6. Open an intensity file from intensity analysis by selecting "File→Open file" in the IGB menu. Open single- or multiple

(by holding shift key)-intensity files and then click "open" to finish. The intensity analysis result is loaded into IGB and visible above the Refseq annotation track (see Note 20).

7. By default, the background of IGB is in black and the annotations are in green. For publication or printing purpose, the background and signal color can be changed. Go to "Graph Adjuster" tab in the lower window to change the desired color.

8. IGB does not filter the intensity data. It has to be manually filtered by value or by percentile. Key in "0%" under "Min" and "95%" under "Max" in the "Graph Adjuster" tab to analyze the top 5% percentile of data*. The selection can be also adjusted dynamically by using the adjacent slide bar (see Note 21). An example showing expression profile output for heat-shock protein 1 (Hspe1) is illustrated in Fig. 2.

3.8.5. Calculate Intervals

1. Define the thresholds for significant transcribed regions in IGB by highlighting the result track from intensity analysis (see Note 22).

2. Select "Graph Adjuster" tab in the lower window of IGB and click "Thresholding" to open graph thresholds option.

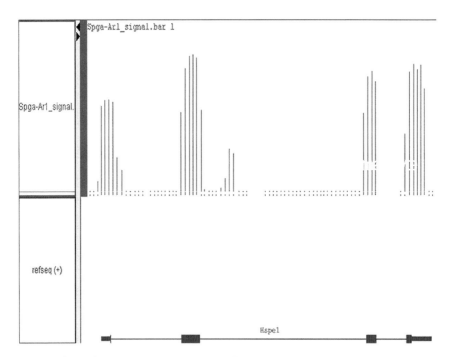

Fig. 2. Expression profiling of mouse heat-shock protein 1 (*Hspe1*) in mouse spermatogonia. The top panel represents filtered intensity signal from a triplicate array set cutoff at 95% percentile. The lower panel represents Refseq annotation of *Hspe1* from Genbank. Note that the intensity signal aligns perfectly with the exon annotation of *Hspe1*. The tiling array signals also indicate that a genomic region between exon 2 and 3 of *Hspe1* is expressed, which was not annotated by any public database.

3. Turn "Visibility" to "On" and select the threshold value at 95% percentile. Put 230 in Max Gap and 60 in Min Run windows. Regions that meet the threshold criteria are visible under the signal track as solid blocks (see Note 23).

4. To create an independent track, click "Make Track" in the graph thresholds window.

5. Right click the track and select "Save Annotations" to export the track data in .bed file format. The bed files can be viewed in IGB or UCSC genome browser.

6. Alternatively, the intervals can be calculated in TAS, select "Edit→defaults," open "Interval Analysis" tab, and put the corresponding numbers from step 3 in each field. Then, go to "Analysis" option in the menu and select "Intervals" to open the intensity files to be analyzed; press "Open" to start intervals analysis.

4. Notes

1. Do not allow the magnetic beads to dry out. Dried beads reduce the recovery of RNA. Resuspend the beads immediately.

2. To check if rRNA depletion is efficient, load 1 μl of the purified RNA to an RNA 6000 Nano LabChip for RNA profile analysis.

3. Input RNA quality is critical for the overall success of the experiment. The quality and quantities of cRNA generated are very dependent on the $O.D_{260/280}$ ratio and amount of input total RNA. The intensity of signals is sensitive to artifacts contained in the poor-quality samples. According to our experience, we recommend the input total RNA quality having an $O.D_{260/280}$ ratio of at least 2. Three micrograms of total RNA is sufficient for up to nine GeneChip hybridization reactions. To study nonpolyadenylated transcripts or transcripts unique to cytosol or nuclear fraction, additional separation protocols are required before proceeding to cDNA synthesis step.

4. Use SpeedVac to reduce the volume of RNA sample. Working T7 primer is diluted in 1:5 from stock.

5. Make sure that the total cRNA is equal to or above 21 μg. Poor amplification efficiency or RNA input quality could lead to low cRNA yield.

6. Details on gel-shift assay can be found in Appendix A of GeneChip Whole Transcript Double-Stranded Target Assay Manual.

7. Do not discard the remaining cocktail mix. The mix is to be reused to hybridize chips up to three rounds. Store the cocktail at −20°C.

8. Refer to fluidic station operation manual for operation procedure.

9. The scanning takes about 30 min for each chip. Refer to the GCOS operation manual for details on setting up experiment.

10. In most cases, TAS is good for analyzing a simple sample setup like single transcriptome analysis discussed in this chapter. However, it is important to note that TAS has certain limitations. For example, TAS can be used to compare paired samples only and the output file does not include any genomic annotations. To perform multisample comparisons and generate mapped annotation data to the significant intervals, third-party software are required, such as Partek Genomic Suite (http://www.partek.com) or TileMapper (http://www.nichd.nih.gov/TileMapper).

11. Biological instead of technical replicates are highly recommended. To minimize batch-to-batch variations, all replicates should be processed and analyzed together if possible.

12. Naming the TAG file in format like (sample name)-(date)-(Chip number) (e.g., Spga-010309-A1) enables better tracking and prevents confusion in subsequent data handling on the samples.

13. Two normalization methods are available in TAS. They are linear scaling and quantile normalization. In linear scaling, all array samples were scaled to the same value by multiplying signal intensities of each feature in an array by a factor. For quantile normalization, the signal intensity distribution is adjusted equally so that the median signal intensities are equal across the experiments.

14. To lower the noise and susceptibility to outliers, the intensity signal and p-values are calculated based on a cluster of oligonucleotide probes (data points), not by individual calculation. In TAS, such cluster of data points is known as window, which is determined by genomic distance (bandwidth) from the probe set of interest.

15. The optimal size of bandwidth depends on the nature of biological samples and experimental application. For transcriptome analysis, we determined the windows based on average exon size of a transcript. Although a larger bandwidth has more statistical power, it could result in diluted signal values. It is important to test against known controls to come up with a bandwidth size that gives the best data representation.

16. Signal calculation in one sample analysis is based on Hodges–Lehmann estimator defined by the bandwidth size. The median of all pairwise averages derived from the differences of PM and MM signals in every window is calculated. Signals are expressed in log base 2 scale. *P*-values are determined by Wilconxon signed rank test using PM and MM probe value differences within bandwidth area from all replicates. The test is based on the null hypothesis that there is no signal difference between PM and MM probes.

17. The high-resolution data created by tiling microarray is best visualized in genome browser rather than in spreadsheet (text) format. IGB can be used for viewing whole transcriptome intensity analysis result files.

18. IGB can be started from Web or through local installation. To run IGB from Web, make sure that Java Web Start (www.java.com) is installed. IGB requires java at least version 1.4.2. However, version 1.5 is recommended and any more recent version can also be used. The minimum memory requirement for IGB alone is 256 mb of system memory; therefore, a system of at least 512 mb is required. However, we recommend to run IGB in a system with at least 2 gb of system memory. To load the intensity data from all the chips in triplicate setting, start IGB with 1.5 gb memory requirement. Use of lower memory requirement options causes slow down or program crash.

19. Other than known gene/position search, IGB also offers pattern search functionality. Regular expression pattern search is very useful for looking for specific sequence features or finding instances of a sequence containing unknowns. A complete list of regular expression syntax can be found at http://www.java.sun.com/j2se/1.5.0/docs/api/java/util/regex/Pattern.html.

20. In addition to Refseq annotation, IGB also provides access to other annotation information, which can be retrieved from NetAffx or DAS/2 server.

21. The percentile cutoff is very subjective to application. A low percentile cutoff gives more data point at the cost of false positivity. High-percentile cutoff gives better success rate in downstream validation at the cost of data coverage.

22. Interval analysis is performed based on intensity analysis results (bar or chp files). The goal is to define regions of transcripts in the genome based on the signal or *p*-values from intensity analysis. These significant transcribed regions known as transcribed fragments (Transfrags) are revealed as blocks in IGB.

23. Max_gap is defined as maximum distance between positive positions while Min_run is defined as minimum length of positive region. The values for these two parameters are dependent on type of application and adjustment based on known positive controls.

Acknowledgments

This work was supported by the Intramural Research Program of the National Institutes of Health (NIH), Eunice Kennedy Shriver National Institute of Child Health and Human Development.

References

1. Schultz N, Hamra FK, Garbers DL (2003) A multitude of genes expressed solely in meiotic or postmeiotic spermatogenic cells offers a myriad of contraceptive targets. Proc Natl Acad Sci USA 100:12201–12206.

2. Shima JE, McLean DJ, McCarrey JR, Griswold MD (2004) The murine testicular transcriptome: characterizing gene expression in the testis during the progression of spermatogenesis. Biol Reprod 71:319–330.

3. Johnston DS, Wright WW, Dicandeloro P, Wilson E, Kopf GS, Jelinsky SA (2008) Stage-specific gene expression is a fundamental characteristic of rat spermatogenic cells and Sertoli cells. Proc Natl Acad Sci USA 105:8315–8320.

4. Lee TL, Li Y, Alba D, Vong QP, Wu SM, Baxendale V, Rennert OM, Lau YF, Chan WY (2009) Developmental staging of male murine embryonic gonad by SAGE analysis. J Genet Genomics 36:215–227.

5. Lee TL, Li Y, Cheung HH, Claus J, Singh S, Sastry C, Rennert OM, Lau YF, Chan WY (2010) GonadSAGE: a comprehensive SAGE database for transcript discovery on male embryonic gonad development. Bioinformatics 26:585–586.

6. Lee TL, Pang AL, Rennert OM, Chan WY (2009) Genomic landscape of developing male germ cells. Birth Defects Res C Embryo Today 87:43–63.

7. Wrobel G, Primig M (2005) Mammalian male germ cells are fertile ground for expression profiling of sexual reproduction. Reproduction 129:1–7.

8. Chalmel F, Rolland AD, Niederhauser-Wiederkehr C, Chung SS, Demougin P, Gattiker A, Moore J, Patard JJ, Wolgemuth DJ, Jegou B et al (2007) The conserved transcriptome in human and rodent male gametogenesis. Proc Natl Acad Sci USA 104:8346–8351.

9. Pang AL, Taylor HC, Johnson W, Alexander S, Chen Y, Su YA, Li X, Ravindranath N, Dym M, Rennert OM et al (2003) Identification of differentially expressed genes in mouse spermatogenesis. J Androl 24:899–911.

10. Bertone P, Gerstein M, Snyder M (2005) Applications of DNA tiling arrays to experimental genome annotation and regulatory pathway discovery. Chromosome Res 13:259–274.

11. Johnson JM, Edwards S, Shoemaker D, Schadt EE (2005) Dark matter in the genome: evidence of widespread transcription detected by microarray tiling experiments. Trends Genet 21:93–102.

12. Mockler TC, Chan S, Sundaresan A, Chen H, Jacobsen SE, Ecker JR (2005) Applications of DNA tiling arrays for whole-genome analysis. Genomics 85:1–15.

13. Cowell JK, Hawthorn L (2007) The application of microarray technology to the analysis of the cancer genome. Curr Mol Med 7:103–120.

14. Yazaki J, Gregory BD, Ecker JR (2007) Mapping the genome landscape using tiling array technology. Curr Opin Plant Biol 10:534–542.

15. Yagi S, Hirabayashi K, Sato S, Li W, Takahashi Y, Hirakawa T, Wu G, Hattori N, Ohgane J, Tanaka S et al (2008) DNA methylation profile of tissue-dependent and differentially methylated regions (T-DMRs) in mouse promoter regions demonstrating tissue-specific gene expression. Genome Res 18:1969–78.

Chapter 8

Biochemical Characterization of a Testis-Predominant Isoform of N-Alpha Acetyltransferase

Alan Lap-Yin Pang

Abstract

N-alpha protein acetylation, catalyzed by N-alpha acetyltransferase complex, is a common protein modification process in eukaryotic cells. Despite its widespread occurrence, the biological significance of this modification process is still unclear. We recently discovered a novel testis-predominant isoform of the catalytic subunit of the enzyme complex. Here, we describe the biochemical characterization of this testis-predominant N-alpha acetyltransferase complex, which includes protein–protein interaction study by co-immunoprecipitation experiment and functional study by N-alpha acetyltransferase assay.

Key words: Testis, Male germ cells, N-alpha acetyltransferase, Arrest defective 1, Co-immunoprecipitation, Protein–protein interaction

1. Introduction

N-alpha protein acetylation is one of the most common protein modification processes occurring on eukaryotic proteins. Over 80% of mammalian proteins undergo this modification (1, 2). This cotranslational modification process is catalyzed by N-alpha-acetyltransferase (NAT). Among the known forms of NAT, the most studied NatA complex consists of the catalytic subunit ARD1A (Naa10p) and the auxiliary subunit NAT1 (Naa15p). Recent reports suggest that ARD1A alone contributes to tumorigenesis and the NAT1–ARD1A complex regulates neuronal dendritic development (3–5). Nevertheless, the biological significance of N-alpha protein acetylation remains unknown.

A novel isoform of ARD1A called ARD1B (Naa11p) was identified in both human (6) and mouse (7). *ARD1B* transcripts are detected predominantly in the mouse testis; the expression of

Wai-Yee Chan and Le Ann Blomberg (eds.), *Germline Development: Methods and Protocols*, Methods in Molecular Biology, vol. 825, DOI 10.1007/978-1-61779-436-0_8, © Springer Science+Business Media, LLC 2012

ARD1B is upregulated during male meiosis, but its translation is repressed until postmeiotic spermatids develop (7). The testis-predominant expression pattern of *ARD1B* suggests that N-alpha protein acetylation may be crucial for male germ cell development. The high degree of sequence homology between ARD1A and ARD1B implies that they may be functionally similar to each other. We have examined whether ARD1B is a bona fide catalytic subunit of NAT in the testis by studying if ARD1B can form complex with NAT1 (through co-immunoprecipitation experiment) and if the NAT1–ARD1B complex can catalyze the transfer of acetyl group to a synthetic human adrenocorticotropic hormone (ACTH 1–24) peptide (through NAT assay).

2. Materials

2.1. Polymerase Chain Reaction Cloning

1. *Pfu* Turbo DNA polymerase (Stratagene).
2. Gene-specific PCR primers (Operon).
3. Zero-Blunt pCR4-TOPO vector (Invitrogen).
4. *Kpn* I restriction enzyme and T4 DNA ligase (Fermentas).
5. phCMV3 expression vector (Genlantis).
6. Wizard SV Gel and PCR Clean up system (Promega).
7. Plasmid Midi kit (Qiagen).
8. SeaKem GTG Agarose ("low-melting agarose"; certified for the recovery of nucleic acids ≥1 kb) (Cambrex Bio Science Rockland, Inc.).
9. 50× Tris–acetate buffer: 242 g of Tris base, 57.1 mL of glacial acetic acid, and 100 mL of 0.5 M ethylenediamine tetraacetic acid (EDTA), pH 8.0. Dilute to 1× concentration with distilled deionized water before use (see Note 1).

2.2. Cell Culture and Transfection

1. Chinese hamster ovary (CHO)-K1 cells (American Type Culture Collection (ATCC)).
2. F-12K medium supplemented with 10% fetal bovine serum and 1% antibiotic–antimycotic solution (Mediatech).
3. 0.25% solution of trypsin (with EDTA and phenol red) and 1× phosphate buffered saline (PBS) (Invitrogen).
4. Hank's balanced salt solution (HBSS) (Mediatech).
5. FuGENE6 transfection reagent (Roche Applied Bioscience). Equilibrate reagent to room temperature before use.
6. Cell lysis buffer: 50 mM Tris–HCl, pH 7.6, 150 mM NaCl, and 1% Nonidet P-40. Chilled and supplemented with 1× Halt protease inhibitor cocktail and 5 mM EDTA (Pierce) before use.

2.3. Co-immuno precipitation

1. Anti-rabbit Ig IP beads (eBioscience).

2. Rabbit-anti-HA antibody SG77 (Zymed/Invitrogen).

3. Normal rabbit IgG (Santa Cruz Biotechnology).

4. 2× sodium dodecyl sulfate (SDS) protein gel loading solution (Quality Biological): Aliquoted and supplemented with 5% (v/v) 2-mercaptoethanol before use. Discard reconstituted aliquot if left unused for more than 2 weeks.

2.4. Sodium Dodecyl Sulfate-Polyacrylamide Gel Electrophoresis

1. Bicinchoninic acid (BCA) protein assay kit (Pierce).

2. 30% (w/v) Acrylamide:0.8% (w/v) Bis-acrylamide solution mix (UltraPure ProtoGel, National Diagnostics).

3. 1.5 M Tris solution, pH 8.8.

4. 1 M Tris solution, pH 6.8.

5. 10% (w/v) SDS solution.

6. 10% (w/v) ammonium persulfate (BioRad) solution. Freeze aliquots at $-20°C$ and thaw working aliquot one at a time. No adverse effect on gel polymerization was observed with the use of ammonium persulfate solution stored at 4°C for at least 2 weeks.

7. N,N,N',N'-Tetramethylethylenediamine (TEMED) (Invitrogen).

8. 10× SDS-polyacrylamide gel electrophoresis (PAGE) buffer (BioRad): 250 mM Tris, 1.92 M glycine, and 1% SDS, pH 8.3. Dilute to 1× concentration with distilled deionized water before use.

9. Pre-stained SDS-PAGE standards (low range) (BioRad).

10. Grade 3MM Chr chromatography paper (Whatman International Ltd.).

2.5. Western Blotting

1. Mouse anti-myc antibody (Invitrogen).

2. Rabbit anti-HA antibody SG77 (Zymed/Invitrogen).

3. Horseradish peroxidase (HRP)-conjugated goat anti-mouse IgG (BioRad).

4. Chicken anti-myc antibody and HRP-conjugated goat polyclonal to chicken IgY H&L (Abcam).

5. Rabbit IgG TrueBlot (eBioscience).

6. SuperSignal West Pico chemiluminescent substrate (store solutions at 4°C) and Restore western blot stripping buffer (both from Pierce).

7. Polyvinylidene fluoride (PVDF) membrane, 0.2 µm (BioRad).

8. Protein transfer buffer: 5.82 g/L Tris base, 2.93 g/L glycine, 3.75 mL 10% (w/v) SDS solution, and 200 mL absolute methanol. Final pH of solution should be around 9.0–9.4.

9. Blocking solution: 1× PBS, 0.05% Tween-20, and 5% nonfat dried milk. Stir continuously at room temperature for more than 1 h before use. Store solution at 4°C.

10. Washing solution (1× PBST): 1× PBS and 0.05% Tween-20. Store solution at 4°C.

2.6. NAT Assay

1. TNT T7 Quick Coupled Transcription/Translation System (Promega).

2. Nuclease-free water (Ambion).

3. 0.2 M potassium phosphate buffer, pH 8.1.

4. (^{3}H) acetyl-CoA (152 GBq/mmol) (Amersham Biosciences).

5. Synthetic human ACTH (1–24) (Calbiochem): Reconstituted to 0.5 mM with 5% acetic acid.

6. SP Sepharose (Sigma). Storage buffer is exchanged three times with 0.5 M acetic acid before use.

7. Absolute methanol.

8. 1× Tris-buffered saline (TBS): 50 mM Tris–HCl, pH 7.6, and 150 mM NaCl.

3. Methods

3.1. Preparation of Expression Vectors

1. Expression vectors for myc-tagged NAT1 and HA-tagged ARD1A (CS2 + MT-mNAT1 and CS2 + mARD1/HA) are constructed previously (8). All expression vectors are propagated by bacterial transformation and extracted using a Plasmid Midi kit.

2. To express a C-terminal hemagglutinin (HA) epitope-tagged ARD1B protein, the *ARD1B* open reading frame (ORF) is amplified from mouse pachytene spermatocyte cDNA using high-fidelity DNA polymerase, such as *Pfu* Turbo DNA polymerase. Primers flanking the *ARD1B* ORF are designed to include an extra hexameric sequence (TAGGGC) followed by *Kpn*I restriction site (GGTACC) at their 5′ ends to facilitate DNA subcloning procedure (see Note 2).

3. The PCR product is cloned into ZeroBlunt pCR4-TOPO vector, propagated by bacterial transformation, and verified by DNA sequencing. Correct clones of vectors are digested with *Kpn* I to release the *ARD1B* ORF. After electrophoresis on low-melting agarose gel (see Note 3), the *ARD1B* ORF is cut from the gel and extracted using the Wizard SV Gel and PCR Cleanup system. The phCMV3 vector is linearized with *Kpn* I, purified using the same system, and ligated with the *ARD1B* ORF in the presence of T4 DNA ligase. The recombinant plasmid, phCMV3-ARD1B/HA, is also propagated by bacterial transformation.

3.2. Cell Transfection for Co-immuno-precipitation

1. CHO-K1 cells are maintained in T75 culture flasks or 100-mm culture dishes in F-12K medium supplemented with 10% fetal bovine serum and 1% antibiotic–antimycotic solution (working culture medium). At 90% confluence, cells are subcultured by trypsinization and maintained at 10–20% confluence (see Note 4).

2. For transfection experiments, CHO-K1 cells are trypsinized and counted using a hemocytometer. In each well of a 6-well culture plate, 1×10^5 cells are seeded (see Note 5).

3. Co-transfection of CHO-K1 cells with myc-NAT1 and ARD1A-HA or ARD1B-HA expression vectors is performed using FuGENE6 transfection reagent. For transfection of cells seeded in one well, 97 µL of plain F12-K medium (at room temperature) is first added into a sterile Eppendorf tube, followed by 3 µL of FuGENE6 reagent. The mixture is vortexed for 1 s and incubated at room temperature for 5 min. Plasmid DNA (500 ng each per well) is added to the mixture, vortexed for 1 s, and incubated at room temperature for 30 min to allow the formation of DNA–transfectant complex. The whole mixture is then added to the CHO-K1 cells in a dropwise manner. The culture plate is slightly rocked to ensure even distribution of content, and incubated at 37°C with 5% CO_2 (see Note 6).

4. Forty-eight hours later, cells are washed twice with cold HBSS or PBS and lysed with chilled cell lysis buffer (1 mL per well). Viscosity of samples is reduced by passing the cell lysates through a 23-G needle ten times, and the lysates are incubated on ice for 30 min with occasional mixing. Soluble fractions are harvested by centrifugation at $18,000 \times g$ at 4°C for 15 min. Protein content is determined by BCA protein assay.

5. In each co-immunoprecipitation experiment, 500 µg of cell lysate is first precleared by incubating with 25 µL (packed volume) of anti-rabbit Ig IP beads on ice for 30 min with occasional mixing. The supernatant is transferred to a new Eppendorf tube after spinning at $18,000 \times g$ at 4°C for 5 min. Five microgram of rabbit-anti-HA antibody or normal rabbit IgG is added to the precleared lysate and incubated on ice for 2 h. Subsequently, 25 µL (packed volume) of freshly washed anti-rabbit Ig IP beads is added and incubation is continued at 4°C overnight with constant agitation (see Note 7).

6. After overnight incubation, the immunoprecipitate is harvested by spinning at $18,000 \times g$ at 4°C for 2 min and washed three times with 500 µL of chilled cell lysis buffer. The immunoprecipitate is resuspended in 50 µL of 2× SDS-PAGE loading buffer and heated at 95°C for 5 min before loading to protein gel.

3.3. Sodium Dodecyl Sulfate-Polyacrylamide Gel Electrophoresis

1. The instructions described here are based on the use of the Mini PROTEAN 3 Cell from BioRad (see Note 8). A detailed recipe for SDS-PAGE gel preparation can be found in Sambrook et al. (9).

2. For a single 10% resolving gel with a 1.0-mm thickness, prepare the following gel solution (5 mL) in a 50-mL E-flask with each component added accordingly: 1.9 mL of distilled deionized water, 1.3 mL of 1.5 M Tris (pH 8.8), 1.7 mL of 30% acrylamide mix, 50 μL of 10% SDS solution, 50 μL of 10% ammonium persulfate solution, and 2 μL of TEMED. Once the TEMED is added, swirl the flask gently and dispense ~4.5 mL of the gel solution into the gel cassette assembly. The gel surface is then overlaid with ~0.5 mL of isopropanol and gel polymerization is allowed at room temperature for at least 20 min (see Note 9).

3. Remove isopropanol and rinse the gel top with distilled deionized water six times. After the final rinse, leave enough water to cover the gel top to avoid drying.

4. In another 50-mL E-flask, prepare the 5% stacking gel solution (2 mL) by mixing the following components accordingly: 1.4 mL of distilled deionized water, 0.25 mL of 1 M Tris (pH 6.8), 0.33 mL of 30% acrylamide mix, 20 μL of 10% SDS solution, 20 μL of 10% ammonium persulfate solution, and 2 μL of TEMED. Once the TEMED is added, swirl the flask gently and quickly remove the water overlaying gel top with filter paper. Dispense ~1.5 mL of stacking gel solution to the gel cassette and insert the comb without delay. The gel is allowed to polymerize at room temperature for at least 20 min.

5. Mark the bottom of the wells on the glass plates of gel cassettes with a water-insoluble marker.

6. When the stacking gel is set, assemble the gel cassette with the electrode assembly and clamping frame; insert the whole chamber assembly into buffer tank. Fill the inner chamber with fresh 1× SDS-PAGE buffer to the top and observe if buffer leaks from the bottom of the chamber assembly.

7. Carefully remove the comb and flush the wells gently by pipetting with 1× SDS-PAGE buffer. Fill the outer chamber with the same buffer.

8. The heat-denatured immunoprecipitate samples are centrifuged at $18,000 \times g$ at room temperature for 3 min. Carefully pipette 20 μL of the supernatant (without disturbing the agarose pellet) and load into the well. Remember to leave a well for the prestained SDS-PAGE standards (5–10 μL).

9. Close the assembly with the lid and connect to a power supply. The gel is run at 120 V (constant voltage) for ~1 h. The samples

appear to be packed into a blue line through the stacking gel but are resolved in the resolving gel (judging by the separation of protein bands from the SDS-PAGE standards). Stop running when the dye front is ~5 mm away from the lower edge of glass plate. Disassemble the gel cassette and proceed to western blotting procedure.

3.4. Western Blotting

1. The instruction described here is based on the use of Mini Trans-Blot Cell from BioRad. Proteins resolved in SDS-PAGE are electrotransferred to PVDF membrane.

2. Prepare the cooling unit by packing with ice and distilled deionized water and keeping it at −80°C for ~1 h. Prechill the transfer buffer at 4°C ahead of time. When SDS-PAGE is finished, trim the stacking gel by a plastic spatula. Equilibrate the resolving gel with at least 50 mL of chilled transfer buffer in a small plastic container; gently shake the gel at room temperature for at least 15 min.

3. In the presence of methanol, the gel shrinks. Measure the dimensions of the gel with a ruler ~5 min after immersing the gel in transfer buffer. Use clean scissors to cut a piece of PVDF membrane with the same dimension as the shrunken gel. Activate the membrane by wetting it with absolute methanol in a clean plastic container. The membrane should turn from white to semitransparent once "activated." Take the membrane out with clean flat-end forceps and immerse it in ~50 mL of chilled transfer buffer in another small plastic container. Shake gently at room temperature for at least 15 min to equilibrate the membrane (see Note 10).

4. Assemble the gel-membrane "sandwich": soak six pieces of filter paper (cut to the same dimension as the PVDF membrane) and two filter pads in transfer buffer in a clean container. Place the gel holder cassette on flat surface with the clear side down. Place one wet filter pad on the clear side of the cassette and stack with three pieces of wet filter paper. Roll out any trapped air bubble with a stripette every time a piece of wet filter paper is overlaid. Place the equilibrated PVDF membrane on the filter paper stack, followed by the equilibrated gel. Stack the remaining pieces of wet filter paper and subsequently the other wet filter pad.

5. Lock the "sandwich" in the gel holder cassette by closing the white latch. Insert the cassette into electrode module with the grey side of the cassette facing the black side of the module (the clear side of the cassette should face the red side of the module). Insert the whole assembly into buffer tank with the frozen cooling unit. Completely fill the buffer tank with chilled transfer buffer. Place a magnetic stir bar in the tank and keep the buffer stirred on a stirring plate.

6. Complete the assembly by inserting the lid and connecting it to a power supply. Protein transfer takes place at 100 V for 1 h.

7. When the transfer process is finished, disassemble the transfer setup and carefully remove the gel from the membrane. Pencil mark the position of the SDS-PAGE standards' bands immediately and quickly immerse the membrane in blocking solution. Use at least 0.3 mL of blocking solution per 1 cm^2 of membrane. Shake the membrane gently on a shaking platform at room temperature for 1 h.

8. Replace the blocking solution with equal volume of fresh blocking solution. Primary antibody solution is added directly to the blocking solution at a dilution of 1:5,000. The membrane is then shaken gently at 4°C overnight. Primary antibodies used include anti-myc antibody produced from mice or chicken and anti-HA antibody produced from rabbits.

9. After the overnight incubation, discard the primary antibody solution and rinse the membrane once with 2 volumes (according to the volume of blocking solution used) of washing solution. Afterward, wash the membrane three times for 5 min each with 2 volumes of washing solution.

10. Discard washing solution and replace with one volume of fresh blocking solution. The secondary antibody solution is added directly to the blocking solution and mixed by gentle swirling. Shake the membrane gently at room temperature for 1 h. Secondary antibodies used must match the species in which the primary antibody was generated, which include HRP-conjugated goat anti-mouse IgG and HRP-conjugated goat anti-chicken IgY (both used at 1:10,000) and Rabbit IgG TrueBlot™ (1:1,000) (see Note 11).

11. Discard the secondary antibody solution. Rinse and wash the membrane as described in step 9.

12. During the final wash, mix 1 mL of peroxide solution and 1 mL of luminol solution from the SuperSignal West Pico chemiluminescent substrate kit in a clean plastic container. Remove the membrane from the washing solution with forceps and overlay it, with the side containing the blotted proteins facing downward, onto the chemiluminescent substrate mix. Remove any trapped air bubbles and incubate at room temperature for 5 min.

13. Embed the membrane between two pieces of transparency. Seal the openings with masking tapes. Expose the membrane to X-ray film in a film cassette. Determine the optimal exposure time by exposing films for different duration of time. A representative result of western blotting after co-immunoprecipitation is illustrated in Fig. 1a.

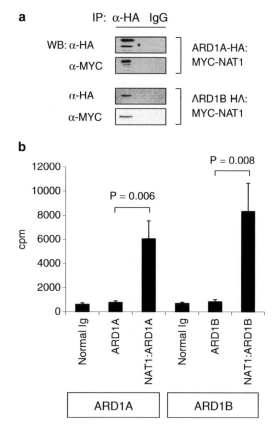

Fig. 1. ARD1B interacts with NAT1 to constitute a functional NAT. (**a**) HA-tagged ARD1B (ARD1B-HA) and MYC-tagged NAT1 (MYC-NAT1) expression vectors were transiently transfected into CHO-K1 cells. Immunoprecipitation (IP) was performed using rabbit anti-HA antibody and normal rabbit IgG, followed by western blotting (WB) with anti-HA and anti-MYC antibodies. The presence of ARD1B-HA and MYC-NAT1 in the same immunoprecipitate indicates a physical interaction between the two proteins. In the control experiment, ARD1A-HA was used instead of ARD1B-HA. The immunoreactive band denoted by the asterisk represents nonspecific signal or ARD1A-HA proteins with partially truncated N-termini. (**b**) NAT assay. In vitro translated proteins were allowed to interact before enriching by immunoprecipitation with rabbit anti-HA antibody or normal rabbit IgG. In normal IgG reactions, ARD1A-HA and ARD1B-HA were separately immunoprecipitated by normal IgG for NAT assay. In ARD1A or ARD1B reactions, ARD1A-HA and ARD1B-HA alone were separately immunoprecipitated by anti-HA antibody for NAT assay. In NAT1:ARD1A and NAT1:ARD1B reactions, the said protein complexes were separately immunoprecipitated by anti-HA antibody for NAT assay. The cpm readings indicate the extent of (^{3}H) acetyl group incorporation to human ACTH 1–24 peptide in the presence of different immunoprecipitates. Figure was originally published in Pang et al. (7).

3.5. Stripping and Reprobing Protein Blots

1. After capturing the optimal exposure level on films, the positions of SDS-PAGE standards' bands are marked on the film by overlaying them onto the transparency enclosing the membrane.

2. The membrane is then taken out with forceps and rinsed once in washing solution. After shaking off the excess liquid, the

membrane is transferred to 8 mL of Restore western blot stripping buffer and shake at room temperature for 15 min. Afterward, the membrane is rinsed once in washing solution and then blocked in blocking solution at room temperature for 15 min.

3. The membrane is now ready for probing with another primary antibody and the detection of immunoreactivity follows the same procedure as mentioned above.

3.6. NAT Assay

1. To avoid interference from endogenous NAT subunits in mammalian cell lines, the enzyme subunits used in NAT assay are prepared by in vitro translation.

2. NAT1, ARD1A-HA, and ARD1B-HA proteins are in vitro translated from pT7-mNAT1 (8), pBS-ARD1/HA (8), and phCMV3-ARD1B/HA expression vectors, respectively, using TNT T7 Quick Coupled Transcription/Translation System. In each translation reaction, 40 µL of TNT T7 Quick Master Mix, 1 µg of plasmid DNA (prepared using Plasmid Midi kit), and 1 µL of methionine (1 mM) are added into a PCR tube and the final reaction volume is made up to 50 µL with nuclease-free water. The reactions are incubated in a thermocycler at 30°C for 90 min.

3. To reconstitute the NAT complex, 20 µL of in vitro translated ARD1A-HA or ARD1B-HA is mixed with 100 µL of in vitro translated NAT1 and the reaction volume is made up to 200 µL with 1× PBS. In the control experiment, the in vitro translated NAT1 is skipped.

4. The HA-tagged protein complexes are immunoprecipitated with anti-HA antibody and washed as described in Subheading 3.2.

5. The NAT assay is performed by incubating the immunoprecipitate, which is immobilized on agarose beads, at 37°C for 2 h in a reaction containing 136 µL of 0.2 M potassium phosphate (K_2HPO_4), pH 8.1, 1 µCi of (^3H) acetyl-CoA (152 GBq/mmol), and 10 µL of 0.5 mM human ACTH (1–24) with constant agitation. Afterward, 130 µL of the supernatant is applied to 150 µL of 1:1 (v/v) slurry of SP Sepharose in 0.5 M acetic acid. The resin is washed three times with 1 mL of 0.5 M acetic acid, rinsed once with 300 µL of absolute methanol, and resuspended in 75 µL of 1× TBS for scintillation counting. A representative result is shown in Fig. 1b.

4. Notes

1. Unless otherwise stated, all solutions are prepared in water that has a resistivity of 18.2 MΩ-cm (distilled deionized water).

2. A certain percentage of the regular desalted grade PCR primer does not contain an intact 5′ end owing to incomplete nucleotide incorporation. Therefore, a random hexameric sequence is included in each primer to ensure that the downstream restriction site is preserved.

3. Regular agarose contains a substance that inhibits ligation reaction. Therefore, low-melting agarose gel should be used if the extraction of DNA insert for cloning is anticipated.

4. Trypsinization procedure: Adherent cells are rinsed once with 1–2 volume of 1× PBS after removal of culture medium by aspiration. PBS is then removed, and 2–3 mL of 0.25% trypsin solution with EDTA is added and gently spread over the cells. The cells are incubated at 37°C for ~5 min or until they are all detached from the culture substrate. Trypsinization is stopped by addition of at least 4–6 mL (i.e., two times the volume of trypsin solution used) of working culture medium. A split ratio of 1:6 to 1:10 is recommended. Culture medium and trypsin solution must be warmed to 37°C before use.

5. Cells can be transfected immediately after seeding or they can be incubated overnight at 37°C with 5% CO_2 before transfection.

6. We recommend to prepare a master mix of reaction mix and dispense the required volume of DNA:transfectant complex to each well accordingly if multiple wells of cells are transfected.

7. Anti-rabbit Ig IP beads are washed three times with chilled cell lysis buffer before use. After each washing step, the content is centrifuged at $18,000 \times g$ at 4°C for 2 min and buffer is removed by aspiration. The precleared lysate–antibody reaction mix is used to resuspend the beads to initiate immunoprecipitation.

8. Glass plates for SDS-PAGE gel assembly have to be thoroughly cleaned and rinsed with distilled water and air dried. Before use, 70% ethanol is sprayed onto glass surface and rubbed with Kimwipes in one direction. Make sure that the glass surface is dust free.

9. Atmospheric oxygen would inhibit the gel polymerization process. Either isopropanol or water-saturated butanol can be used to block contact between the gel solution and air. The alcohol layer also helps maintain the gel top flat. The unused gel solution in the E-flask can serve as an indicator of the completion of polymerization process: simply seal the opening with parafilm and tilt the flask.

10. Put on disposable gloves and use flat-end forceps to handle PVDF membranes at any time.

11. Regular HRP-conjugated anti-rabbit IgG would detect the heavy chain (~ 55 kDa) and light chain (~ 23 kDa) of the antibodies used for immunoprecipitation. The use of Rabbit IgG TrueBlot would eliminate this effect.

Acknowledgments

We would like to thank Roderick A. Corriveau (Department of Cell Biology and Anatomy, Louisiana State University) for the expression vectors CS2 + MT-mNAT1, CS2 + mARD1/HA, pT7-mNAT1, and pBSARD1/HA. This work was supported by the Intramural Research Program of *Eunice Kennedy Shriver* National Institute of Child Health and Human Development, National Institutes of Health.

References

1. Driessen HP, de Jong WW, Tesser GI, Bloemendal H (1985) The mechanism of N-terminal acetylation of proteins. CRC Crit Rev Biochem 18:281–325

2. Arnesen T, Van Damme P, Polevoda B, Helsens K, Evjenth R, Colaert N, Varhaug JE, Vandekerckhove J, Lillehaug JR, Sherman F, Gevaert K (2009) Proteomics analyses reveal the evolutionary conservation and divergence of N-terminal acetyltransferases from yeast and humans. Proc Natl Acad Sci USA 106:8157–8162

3. Seo JH, Cha JH, Park JH, Jeong CH, Park ZY, Lee HS, Oh SH, Kang JH, Suh SW, Kim KH, Ha JY, Han SH, Kim SH, Lee JW, Park JA, Jeong JW, Lee KJ, Oh GT, Lee MN, Kwon SW, Lee SK, Chun KH, Lee SJ, Kim KW (2010) Arrest defective 1 autoacetylation is a critical step in its ability to stimulate cancer cell proliferation. Cancer Res 70:4422–4432

4. Lee CF, Ou DS, Lee SB, Chang LH, Lin RK, Li YS, Upadhyay AK, Cheng X, Wang YC, Hsu HS, Hsiao M, Wu CW, Juan LJ (2010) hNaa10p contributes to tumorigenesis by facilitating DNMT1-mediated tumor suppressor gene silencing. J Clin Invest 120:2920–2930

5. Ohkawa N, Sugisaki S, Tokunaga E, Fujitani K, Hayasaka T, Setou M, Inokuchi K (2008) N-acetyltransferase ARD1-NAT1 regulates neuronal dendritic development. Genes Cells 13:1171–1183

6. Arnesen T, Betts MJ, Pendino F, Liberles DA, Anderson D, Caro J, Kong X, Varhaug JE, Lillehaug JR (2006) Characterization of hARD2, a processed hARD1 gene duplicate, encoding a human protein N-alpha-acetyltransferase. BMC Biochem 7:13

7. Pang AL, Peacock S, Johnson W, Bear DH, Rennert OM, Chan WY (2009) Cloning, characterization, and expression analysis of the novel acetyltransferase retrogene Ard1b in the mouse. Biol Reprod 81:302–309

8. Sugiura N, Adams SM, Corriveau RA (2003) An evolutionarily conserved N-terminal acetyltransferase complex associated with neuronal development. J Biol Chem 278: 40113–40120

9. Sambrook J, Fritsch EF, Maniatis T (1989) Molecular cloning – A Laboratory Manual (2nd ed). Cold Spring Harbor Laboratory Press, New York

Chapter 9

Identification of Novel Long Noncoding RNA Transcripts in Male Germ Cells

Tin-Lap Lee, Amy Xiao, and Owen M. Rennert

Abstract

Emerging evidence from these studies suggested that the male germ cell transcriptome is more complex than previously envisioned. In addition to protein-coding genes, the transcriptome also encodes a significant number of nonprotein-coding transcripts. These noncoding (nc) RNAs appear to be involved in a variety of cellular activities, ranging from simple housekeeping to complex regulatory functions. A class of ncRNAs known as long ncRNAs (lncRNAs) were recently shown to be expressed in a developmentally regulated manner during brain and embryonic stem cell development. This protocol aims to predict and identify potential lncRNA candidates using Serial Analysis of Gene Expression (SAGE) data. We also illustrate how to validate the potential lncRNAs by expression analyses using real-time PCR and Northern Blot. Potential lncRNA candidates in male germ cells are identified using our previously established male germ cell SAGE database (GermSAGE).

Key words: Long noncoding RNA, Male germ cells, Development, SAGE

1. Introduction

An important observation emerged from various genome-wide transcriptome studies is that the majority of transcripts do not code for proteins. They are referred to as noncoding RNAs (ncRNAs) (1). Mammalian cells produce thousands of ncRNAs of unknown function. These nonprotein-coding portions of the genome were often considered "junk," but recent research has shown that ncRNAs have a wide range of regulatory functions. Initial discoveries identified small ncRNAs, such as microRNA (miRNA), small interfering RNA (siRNAs), and Piwi-interacting RNA (piRNA); these have been widely reported to function in various regulatory processes, including male germ cell development (2). Recently, a new class of

Wai-Yee Chan and Le Ann Blomberg (eds.), *Germline Development: Methods and Protocols*, Methods in Molecular Biology, vol. 825, DOI 10.1007/978-1-61779-436-0_9, © Springer Science+Business Media, LLC 2012

ncRNAs known as long ncRNAs (lncRNAs), defined as ncRNA species at least 200 nucleotides in length, has also been demonstrated to function in developmental regulations in brain and mouse embryonic stem cells (ESCs) (3, 4). The precise functional role of lncRNA in male germ cell development remains largely unknown. Its functional importance in developmental regulations suggests that lncRNAs may also be indispensable in male germ cell development.

The presence of ncRNAs was functionally implicated in our previous male germ cell transcriptome studies employing Serial Analysis of Gene Expression (SAGE) (5–10). This sequence-based genomic assay provides an unbiased interrogation of all polyadenylated transcripts in male germ cell transcriptome. Up to 75% of tags were not properly annotated in our male germ cell SAGE libraries. Therefore, we hypothesize that a subset of unannotated SAGE tags may be related to ncRNAs. In this protocol, we demonstrate how to identify lncRNAs in spermatids by using the GermSAGE database (11), a publicly available SAGE database for male germ cell transcriptome data.

2. Materials

2.1. Male Germ Cell Samples

1. Type A spermatogonia were isolated from 6-day-old Balb/c mice (university stocks, Georgetown University). Since 6-day-old mice testis do not contain spermatocyte or spermatid, type A spermatogonia are typically >99% pure after separation by differential plating from contaminating Sertoli cells.

2. RNA quality is critical to the overall success of the experiment. Minimize the number of freeze and thaw cycle if possible and make sure that the experiment is performed in an RNAase-free environment.

2.2. Serial Analysis of Gene Expression (Optional)

1. I-SAGE kit module (Invitrogen).
2. I-SAGE Ditag PCR Module (Invitrogen).
3. Cloning kit (Qiagen).
4. ABI 3730xl sequencer.
5. SAGE2000 software.

2.3. Quantitative Real-Time PCR Validation

1. Primers are designed by Primer3 and show no significant homology with other genomic regions by NCBI Blast.
2. Taqman PCR Reagent Kit: AmpliTaqGold DNA polymerase, dUTP, dATP, dCTP, dGTP, 1.2 ml 10× Taqman buffer A, and 2×1.5 ml 25 mM $MgCl_2$.
3. Bovine serum albumin (BSA).

4. 1× Tris–EDTA buffer (10 mM Tris–HCl containing 1 mM EDTA·Na$_2$).

5. ABI PRISM Sequence Detection System.

2.4. Northern Blot Analysis

1. Digoxigenin (DIG) Northern Starter Kit (Roche).

2. DIG Easy Hyb Granules.

3. Anti-DIG-AP, Fab fragments.

4. Chemiluminescent substrate.

5. Actin RNA Probe, DIG labeled.

6. DIG-11-UTP.

7. DIG wash and block buffer set.

8. Nylon membrane, positively charged.

9. Hybridization bags.

10. 20× SSC buffer (pH 7.0).

11. Immunological detection solutions.

12. Stripping buffer:50% deionized formamide, 5% SDS, 50 mM Tris–HCl, adjust to pH 7.5.

Solution	Preparation
Washing buffer	0.1 M maleic acid, 0.15 M NaCl, and 0.3% Tween 20 (pH 7.5)
Maleic acid buffer	0.1 M maleic acid, 0.1 M NaCl (pH 7.5)
Detection buffer	0.1 M Tris–HCl, 0.1 M NaCl (pH 9.5)
1× blocking solution	Dilute 10× stock with maleic acid (1:10)
Antibody solution	Centrifuge anti-DIG-AP for 5 min at 13,750 × g (10,000 rpm) and dilute in blocking solution (1:10,000)

13. Stripping solution (optional): DEPC water, stripping buffer, 20× SSC, and 2× SSC.

2.5. Bioinformatics Tools and Databases

1. SAGE2000 software (Invitrogen).

2. Primer3 (http://frodo.wi.mit.edu/primer3/).

3. NCBI Blast (http://blast.ncbi.nlm.nih.gov).

4. RNAfold (http://rna.tbi.univie.ac.at/cgi-bin/RNAfold.cgi).

5. SAGEmap (http://www.ncbi.nlm.nih.gov/projects/SAGE/).

6. GermSAGE (http://germsage.nichd.nih.gov/).

7. UCSC genome browser (http://genome.ucsc.edu/).

8. Galaxy (http://galaxy.psu.edu/).

9. fRNAdb (http://www.ncrna.org/frnadb/).

3. Methods

3.1. Construction of SAGE Libraries (Optional, see Note 1)

1. Mix 5 µg of total RNA with oligo dT magnetic beads.
2. Synthesize double-strand cDNA.
3. Digest with NlaIII to form one end of the tag.
4. Divide in half and ligate 40-bp adapters containing the recognition sequence for the type-II restriction enzyme BsmF1.
5. Cleave with BsmF1 to form ~50 bp tag (40 bp adaptor/13 bp tag).
6. Fill in 5′ overhangs and ligate to form an ~100-bp ditag.
7. PCR amplify using ditag primers 1 and 2.
8. Cut 40-bp adapters with Nla III to release the 26 bp ditag.
9. Ligate ditags to form concatemers.
10. Clone into pZErO®-1 and sequence.
11. Annotate the tag by SAGEmap.

3.2. Identification of Spermatid-Specific SAGE Tags from GermSAGE

1. The data from SAGE libraries have been deposited in the publicly available GermSAGE database (http://germsage.nichd.nih.gov/). Download the raw data files from the three main germ cell stages (type A spermatogonia, pachytene spermatocytes, round spermatids) from GermSAGE.
2. To screen for abundantly expressed tags, filter the tag count by equal or greater than five.
3. Transcripts represented by the tags in step 2 may be expressed in more than one male germ cell stage. To identify stage-specific expressed tags, apply Venn diagram analysis on the three datasets (see Note 2).
4. Identify genomic coordinates of Sptd tags by BLAST (see Note 3).
5. To increase the reliability of transcript prediction, align the coordinate results with polyadenylation (PolyA) signal data (see Note 4).
6. Identify stand orientation of tags, and calculate the relative distance of polyA signal to the 3′ end of each tag. Tags without PolyA evidence are eliminated.
7. Download mouse long noncoding transcript track data (fRNAdb::uc-ncRNA) from fRNAdb (see Note 5).
8. Import the long noncoding transcript data from fRNAdb and the filtered tag data from step 6 to Galaxy.
9. Use "Operate on Genomic Intervals" function in Galaxy to align both datasets. Set to detect the presence of tags around

500 bp to both ends. Inconsistent orientations between the transcript and the tag are removed.

10. Determine the stability of transcripts by RNAfold by uploading the sequence in FASTA format. Select "minimum free energy (MFE) and partition function" and check "avoid isolated base pairs" under "Fold algorithms and basic options." Check all boxes in the "Output options" section to generate interactive RNA secondary structure plot, RNA secondary structure plots with reliability annotation, and mountain plot.

11. Rank the transcripts based on the distance of matching SAGE tag to the 3′ end of the lncRNA transcripts. This is calculated by subtracting the genomic coordinate at the 3′ end of the transcript from the start position of SAGE tag.

12. Use "Intersect" function of Galaxy to find the association of Sptd lncRNAs to the existing Refseq annotation. Classify the lncRNAs into one of the following categories: (a) Promoter (7), (b) Intronic (3), (c) 3′UTR (8), and (d) Intergenic (6) (see Note 6).

13. Examine the Sptd lncRNA and tag annotation in UCSC genome browser (optional). Sptd lncRNAs with high conversion evidence (MultiZ) score are selected first.

3.3. Validate Stage-Specific Expression by Quantitative Real-Time PCR

1. Design primer sets targeting the predicted lncRNA transcripts by Primer3. Include 18 s rRNA actin or tubulin as reference control (see Note 7).

2. Reverse transcribe RNA into cDNA by using random hexamers as follows:

Reagent	Volume for each reaction
10× Taqman RT buffer	10 µl
25 mM MgCl$_2$	22 µl
Random hexamers (50 ng/ml)	5 µl
500 µM dNTP mix	20 µl
Reverse transcriptase (50 U/µl)	2.5 µl
DEPC water	Variable
Final volume	100 µl

3. Mix and incubate in a PCR thermal cycler as follows: At 25°C for 10 min, 48°C for 30 min, and 95°C for 5 min.

4. Prepare qPCR component as follows:

Reagent	Volume for each reaction
2× SYBR green mix	15 µl
Primers 10 µM (forward)	1 µl
Primers 10 µM (reverse)	1 µl
cDNA template	1 µl
DEPC water	Variable
Final volume	25 µl

5. Mix and incubate in a real-time PCR machine. The conditions for the amplification are as follows: At 50°C for 2 min, 95°C for 10 min followed by 95°C for 15 s, 60°C for 30 s, 72°C for 30 s for 40 cycles, and end the protocol at 72°C for 10 min.

6. Analyze the data according to manufacturer's protocol.

3.4. Confirmation of Predicted ORF by Northern Blots

1. Reverse transcribe input RNA (>1 µg) using a standard reverse transcriptase system.

2. Prepare DNA template from total RNA according to the table below:

Reagent	Volume
DEPC-treated water	Variable
10× expand buffer	5 µl
10 mM d(ACGT)P	1 µl each
Primer (sense)	1 µl
Primer (antisense)	1 µl
Expand high fidelity	0.75 µl
cDNA	2 µl
Final volume	50 µl

3. Incubate the samples in a thermocycler as follows: 94°C for 45 s, 60°C for 45 s, and 72°C for 90 s. Repeat the reaction for 30 cycles.

4. Mix 4 µl of PCR product and 6 µl of DEPC-treated water. Prepare the DIG-labeling mix as follows (see Note 8):

Reagent	Volume
5× labeling mix	4 µl
5× transcription buffer	4 µl
RNA polymerase (SP6, T7, or T3)	2 µl

5. Mix the PCR product and the DIG-labeled mix. Incubate at 42°C or 1 h.

6. Add 2 μl DNA to remove template DNA. Incubate at 37°C for 15 min.

7. Stop the reaction by adding 2 μl of 0.2 M (pH 8.0) EDTA.

8. To determine labeling efficiency, apply 1 μl of labeled probes and controls to the nylon membrane (see Note 9).

9. Fix the probes on membrane by baking at 120°C for 30 min.

10. Wash the membrane with 20 ml of washing buffer and incubate under shaking at room temperature for 2 min (see Note 10).

11. Remove the washing buffer and replace with 10 ml of blocking solution. Incubate for 30 min.

12. Remove the blocking solution and replace with 10 ml of antibody solution. Incubate for 30 min.

13. Remove the antibody solution and wash with 20 ml of washing buffer for 15 min.

14. Place the membrane on a development folder and apply four drops of CDP-star. Incubate at room temperature for 5 min.

15. Expose the membrane to an X-ray film at room temperature for 20 min.

16. Compare the spot intensity of labeling reaction to control probe to evaluate the amount of DIG-labeled product.

17. To prepare for gel electrophoresis, add 20 μl of loading buffer to the RNA probe, denature at 65°C for 15 min, and cool down on ice for 1 min.

18. Run the product on a formaldehyde gel for 2 h (see Note 11).

19. Evaluate the quality of target RNA by staining in 0.25 μg/ml of ethidium bromide briefly. Examine the gel under UV light.

20. Transfer the RNAs using standard transfer protocol (see Note 12). Rinse gel in 20× SSC for 15 min twice.

21. Fix the RNAs to the membrane by UV-cross-linking or baking. For UV cross-linking, place the membrane on a Whatman paper soaked with 2× SSC. For baking, briefly rinse membrane in 2× SSC twice. Bake at 80°C for 2 h (see Note 13).

22. Prepare hybridization solution by reconstituting granules in 64 ml of DEPC-treated water at 37°C.

23. Prewarm 15 ml of DIG Easy Hyb solution to 68°C. Prehybridize the membrane in prewarmed DIG Easy Hyb for 30 min with gentle agitation.

24. Denature DIG-labeled RNA probe (100 ng/ml) at 99°C in a thermal cycler for 5 min. Place the probe on ice immediately.

25. Mix the denatured DIG-labeled RNA probe with prewarmed DIG Easy Hyb (see Note 14).

26. Remove prehybridization solution and replace with hybridization solution. Incubate overnight with gentle agitation.

27. Remove the hybridization mix and wash in 2× SSC with 0.1% SDS at room temperature under constant agitation. Repeat the wash twice.

28. Wash twice with 0.1× SSC in 0.1% SDS prewarmed at 68°C under constant agitation.

29. Rinse the membrane with washing buffer for 5 min.

30. Remove washing buffer and incubate with 100 ml of blocking solution for 30 min.

31. Remove blocking solution and incubate with 50 ml of antibody solution for 30 min.

32. Remove antibody solution and wash with 100 ml of washing buffer for 15 min. Repeat twice.

33. Remove washing buffer and incubate with 100 ml of detection buffer for 5 min.

34. Put the membrane in a hybridization bag, add 1 ml of CDP-star solution to soak the membrane. Incubate at room temperature for 5 min (see Note 15).

35. Put the membrane to an image cassette with an X-ray film. Expose at room temperature for 15–25 min (see Note 16).

3.5. Stripping and Reprobing of RNA Blots (Optional)

Stripping and reprobing allow signal optimization and reuse of the existing membrane without tedious RNA loading and membrane preparation procedure. It is important to note that only membranes without drying at any time during the hybridization and detection procedure could be used.

1. Prewarm water bath to 80°C.

2. Place the membrane in a hybridization bag and rinse membrane in DEPC-treated water.

3. Replace DEPC-treated water with stripping solution and incubate at 80°C for 60 min to remove the DIG-labeled probe.

4. Aspirate the stripping buffer. Wash in 2× SSC for 5 min twice.

5. Prehybridize and hybridize with another probe as mentioned in the previous section. Store membrane in a sealed bag at 4°C if not used immediately.

4. Notes

1. Please refer the outline of a SAGE experiment to manufacturer's protocol. This is optional in this chapter. We applied data from GermSAGE database in this protocol. The instruction for data downloading in GermSAGE is indicated in the help section of the Web site.

2. Venn diagram can be generated by using BioVenn (http://www.cmbi.ru.nl/cdd/biovenn/).

3. Local Blast is more flexible. The window size is set to 7.

4. As transcripts identified by SAGE should be polyadenylated, including nearby polyadenylated signal evidence increases the reliability of prediction. Polyadenylated database can be found at PolyA_DB (http://polya.umdnj.edu/polyadb/).

5. fRNAdb is a database for comprehensive noncoding RNA sequences. fRNAdb::uc-ncRNA track contains unannotated lncRNAs with only one exon. This eliminates the complexity of ORF predication and follow-up validations.

6. The workflow details of Galaxy can be found at http://main.g2.bx.psu.edu/screencast.

7. Parameters to be measured during primer design are as follows: melting temperature, free energy for the formation of hairpins, self-dimerization or heterodimer formation should be below $\Delta G \le -9$ kcal/mol. Verify that the amplicon with values of $\Delta G \le -9$ kcal/mol using the DINAMelt Server (http://dinamelt.bioinfo.rpi.edu) to avoid problematic secondary structures.

8. DIG-labeled single-stranded RNA probes are constructed by in vitro transcription of input template DNA. DNA templates have to be linearized at a restriction site of cloned insert. The transcribed sequence should be between 200- and 1,000-bp long. High concentration of EDTA (>0.1 mM) inhibits transcription reaction.

9. The yield of DIG-labeled RNA is critical for hybridization results. High probe concentration causes significant background while too low concentration causes compromised signals.

10. The volume is for small membrane (3×5 cm), which can be processed in a plastic container or petri dish.

11. Run the gel at 4 V/cm in RNase-free gel for at least 2 h. Overnight run is preferred.

12. The details of transfer procedure can be found at the following URL: http://www.nature.com/nmeth/journal/v2/n12/full/nmeth1205-997.html. Use only positively charged nylon membrane. Usually, gels blotted by capillary transfer in 20× SSC overnight give the best results. Alkali transfer is not recommended.

13. We found that baking the membrane at 80°C gave the most consistent results. Baking at 120°C is another option, but not recommended.

14. The homology between the probe and target RNA is a critical factor for determining hybridization condition. Hybridization

stringency is dependent on temperature. High temperature increases stringency, and low temperature decreases stringency. The optimal temperature is 68°C for DIG-labeled probes.

15. Do not allow membrane to dry and make sure that enough solution covers the blotting area to avoid "halo" effect in developed image. Make sure to spread the substrate evenly to prevent formation of air bubbles over the membrane. Avoid adding too much solution or use large-size hybridization bag. Squeeze out excess solution if necessary. Unevenly soaked membrane results in dark background.

16. Luminescence could last up to 24 h. Therefore, multiple exposure could be made within this period.

Acknowledgments

This work was supported by the Intramural Research Program of the National Institutes of Health (NIH), Eunice Kennedy Shriver National Institute of Child Health and Human Development.

References

1. Mattick JS (2001) Non-coding RNAs: the architects of eukaryotic complexity. EMBO Rep 2:986–991.

2. Mattick JS, Makunin IV (2005) Small regulatory RNAs in mammals. Hum Mol Genet 14 Spec No 1:R121–132.

3. Mercer TR, Dinger ME, Mattick JS (2009) Long non-coding RNAs: insights into functions. Nat Rev Genet 10:155–159.

4. Taft RJ, Pang KC, Mercer TR, Dinger M, Mattick JS (2010) Non-coding RNAs: regulators of disease. J Pathol 220:126–139.

5. Wu SM, Baxendale V, Chen Y, Pang AL, Stitely T, Munson PJ, Leung MY, Ravindranath N, Dym M, Rennert OM et al (2004) Analysis of mouse germ-cell transcriptome at different stages of spermatogenesis by SAGE: biological significance. Genomics 84:971–981.

6. Chan WY, Lee TL, Wu SM, Ruszczyk L, Alba D, Baxendale V, Rennert OM (2006) Transcriptome analyses of male germ cells with serial analysis of gene expression (SAGE). Mol Cell Endocrinol 250:8–19.

7. Chan WY, Wu SM, Ruszczyk L, Law E, Lee TL, Baxendale V, Lap-Yin Pang A, Rennert OM (2006) The complexity of antisense transcription revealed by the study of developing male germ cells. Genomics 87:681–692.

8. Lee TL, Li Y, Alba D, Vong QP, Wu SM, Baxendale V, Rennert OM, Lau YF, Chan WY (2009) Developmental staging of male murine embryonic gonad by SAGE analysis. J Genet Genomics 36:215–227.

9. Lee TL, Li Y, Cheung HH, Claus J, Singh S, Sastry C, Rennert OM, Lau YF, Chan WY (2010) GonadSAGE: a comprehensive SAGE database for transcript discovery on male embryonic gonad development. Bioinformatics 26:585–586.

10. Lee TL, Pang AL, Rennert OM, Chan WY (2009) Genomic landscape of developing male germ cells. Birth Defects Res C Embryo Today 87:43–63.

11. Lee TL, Cheung HH, Claus J, Sastry C, Singh S, Vu L, Rennert O, Chan WY (2009) GermSAGE: a comprehensive SAGE database for transcript discovery on male germ cell development. Nucleic Acids Res 37:D891–897.

Chapter 10

Methylation Profiling Using Methylated DNA Immunoprecipitation and Tiling Array Hybridization

Hoi-Hung Cheung, Tin-Lap Lee, Owen M. Rennert, and Wai-Yee Chan

Abstract

DNA methylation is an important epigenetic modification that regulates development and plays a role in the pathophysiology of many diseases. It is dynamically changed during germline development. Methylated DNA immunoprecipitation (MeDIP) is an efficient, cost-effective method for locus-specific and genome-wide analysis. Methylated DNA fragments are enriched by a 5-methylcytidine-recognizing antibody, therefore allowing the analysis of both CpG and non-CpG methylation. The enriched DNA fragments can be amplified and hybridized to tiling arrays covering CpG islands, promoters, or the entire genome. Comparison of different methylomes permits the discovery of differentially methylated regions that might be important in disease- or tissue-specific expression. Here, we describe an established MeDIP protocol and tiling array hybridization method for profiling methylation of testicular germ cells.

Key words: MeDIP, DNA methylation, Tiling arrays

1. Introduction

DNA methylation plays a fundamental role in epigenetic regulation of gene expression and genomic imprinting. In germ cells, methylation is under tight control both to temporally and spatially regulating the expression of genes important for development. X chromosome inactivation and imprinting marks in the germ line are mediated by methylation. Various methods have been developed for analysis of DNA methylation, including bisulfite-based PCR or sequencing, restriction enzyme-based analysis of methylated sites, and immunoprecipitation of methylated fragments. Methylated DNA immunoprecipitation (MeDIP) was first described by Weber et al. (1) and subsequently used for genome-wide analysis of plant and mammalian methylomes (2, 3). MeDIP products can be used for PCR analysis of an individual locus or

Wai-Yee Chan and Le Ann Blomberg (eds.), *Germline Development: Methods and Protocols*, Methods in Molecular Biology, vol. 825, DOI 10.1007/978-1-61779-436-0_10, © Springer Science+Business Media, LLC 2012

hybridized to tiling arrays (MeDIP-chip) for genome-wide analysis. The MeDIP-chip provides cost-effective and high-resolution profiling of the methylome. With the advance of high-throughput next-gen sequencing, sequencing of MeDIP product (MeDIP-seq) or direct sequencing of bisulfite-converted DNA has been applied to resolve methylome in single-base resolution (4, 5). However, MeDIP-chip is still attractive for genomic research because of its rapid and lower cost.

For MeDIP analysis, 1–5 μg of genomic DNA is required. If the downstream analysis is whole-genome array hybridization, 5 μg of DNA is necessary. Enrichment of methylated DNA by anti-5-methylcytidine antibody is effective and sensitive; the degree of enrichment is reflective of the degree of methylation. The antibody can bind to both CpG- or non-CpG-methylated sites. Alternatively, recombinant proteins of the methylated DNA-binding domain (MBD) are used to enrich methylated CpG fragments (6). Both antibody and MBD are sensitive and specific; however, a nonspecific IgG control or input DNA should be included to assess background noise during hybridization.

In this chapter, antibody-based MeDIP is described. We provide procedures for amplification, fragmentation, and labeling of immunoprecipitated DNA for hybridization of tiling arrays using the Affymetrix platform.

2. Materials

2.1. Extraction and Sonication of Genomic DNA

1. Any commercial kit for the extraction of high-quality genomic DNA can be utilized. The procedures described here utilize Gentra Puregene Cell Kit (Qiagen) for purification of RNA-free genomic DNA from cultured cells (see Note 1).

2. Sonics® Ultrasonic Processor Model GE505 (Sonics).

3. Sterile 1× PBS (137 mM NaCl, 2.7 mM KCl, 4.3 mM Na_2HPO_4, 1.47 mM KH_2PO_4, pH 7.4).

4. TE buffer (10 mM Tris–HCl, 1 mM EDTA, pH 8.0).

2.2. MeDIP

1. Antibody: Mouse monoclonal anti-5-methylcytidine (5-mC), Clone 33D3 (Eurogentec).

2. Sheep anti-mouse IgG conjugated Dynabeads (Invitrogen).

3. Proteinase K (Roche).

4. Digestion buffer: 50 mM Tris–HCl (pH 8.0), 10 mM EDTA, 0.5% SDS.

5. 1× IP buffer: 10 mM sodium phosphate (pH 7.0), 140 mM NaCl, 0.05% Triton X-100.

6. Phenol:chloroform:isoamyl alcohol (25:24:1, v/v).

7. Magnetic rack (Invitrogen).

2.3. Amplification of MeDIP DNA

1. Sequenase Version 2.0 DNA Polymerase (13 U/μl) (USB).

2. 5× sequenase reaction buffer: 200 mM Tris–HCl (pH 7.5), 100 mM MgCl$_2$, 250 mM NaCl.

3. Sequenase dilution buffer: 10 mM Tris–HCl (pH 7.5), 5 mM DTT, 0.1 mM EDTA.

4. Primer A (200 μM): GTTTCCCAGTCACGGTC(N)$_9$, HPLC purified.

5. Primer B (100 μM) GTTTCCCAGTCACGGTC, HPLC purified.

6. DMSO, BSA, DTT, dNTP, MgCl$_2$ (molecular grade, various sources).

7. 10× PCR buffer: 200 mM Tris–HCl (pH 8.4), 500 mM KCl.

8. dNTP/dUTP mix: 10 mM of dATP, 10 mM of dCTP, 10 mM of dGTP, 8 mM of dTTP, and 2 mM of dUTP.

9. Platinum Taq (5 U/μl) (Invitrogen).

10. Thermocylcer (Applied Biosystems).

11. GeneChip Sample Cleanup Module (Affymetrix) or MinElute Reaction Cleanup Kit (Qiagen).

2.4. Assessment of Amplified Product by qPCR

1. SYBR Green Master Mix (Applied Biosystems).

2. Control primer pairs for amplifying known methylated and unmethylated loci.

3. Realtime PCR machine (Applied Biosystems).

2.5. Tiling Array Hybridization

1. GeneChip WT Double-Stranded DNA Terminal Labeling Kit (Affymetrix).

2. GeneChip Hybridization, Wash and Stain Kit (Affymetrix).

3. GeneChip Human or Mouse Tiling Array 2.0R (Affymetrix).

4. Control Oligonucleotide B2, 3 nM (Affymetrix).

5. GeneChip® Hybridization Oven 640 (Affymetrix).

6. GeneChip Fluidics Station 450 or 400 (Affymetrix).

7. GeneChip Scanner 3000 7G (Affymetrix).

3. Methods

3.1. Extraction of Genomic DNA

1. Trypsinize cells and resuspend cells in PBS. Count cells.

2. Make an aliquot of ~1 million of cells into a 1.5-ml microcentrifuge tube. Centrifuge at 13,000–16,000 × g for 5 s to collect cell pellet.

3. Discard PBS. Lyse cells immediately or freeze at −80°C.

4. Add 300 μl of cell lysis solution. Lyse cells by pipetting up and down ~50–60 times or vigorous vortexing until the solution becomes aqueous.

5. (Optional) Add 1.5 μl of RNase A solution. Mix and incubate at 37°C for 30 min.

6. Add 100 μl protein precipitation solution. Vortex for 20 s. Centrifuge at maximum speed for 1 min.

7. Add 300 μl of isopropanol to a clean 1.5-ml microcentrifuge tube. Transfer the supernatant and mix by inverting 50 times.

8. Centrifuge at maximum speed for 2 min.

9. Discard supernatant. Add 300 μl of 70% ethanol and invert several times to wash the DNA pellet.

10. Centrifuge at maximum speed for 1 min.

11. Discard the supernatant and remove any residual ethanol. Air dry the DNA for ~10–15 min.

12. Add ~50–100 μl of DNA hydration solution. Hydrate the DNA pellet by pipetting up and down.

13. Incubate at 65°C for 1 h.

14. Take 1 μl for determining the concentration at 260, 280, and 320 nm, respectively.

15. DNA concentration $(\mu g/\mu l) = (A_{260} - A_{320}) \times 0.05 \times$ dilution factor.

16. Adjust DNA to ~0.1 μg/μl with TE. Proceed to sonication or store the DNA at −20°C.

3.2. Sonication of Genomic DNA

1. Dilute ~6 μg of genomic DNA in ~400 μl TE buffer in a 1.5-ml microcentrifuge tube.

2. Prechill the DNA by placing the tube on ice for at least 10 min.

3. Set up the sonicator with the following parameters:

   (a) Amplitude: 20%

   (b) Pulser: 15 s ON; 15 s OFF

   (c) Timer: 10 min

4. Put a microcentrifuge tube in a rack. Fix the rack tightly on an ice-water cup (see Note 2).

5. Dip the sonication probe into the microcentrifuge tube such that the probe is just above the bottom.

6. Start sonication.

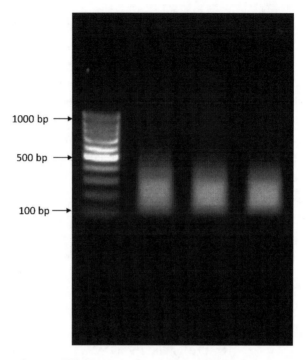

Fig. 1. Genomic DNA fragments after sonication. Mobilities of size markers are *indicated on the left.*

7. After sonication, take 15 µl of the sonicated DNA and check the efficiency of fragmentation by electrophoresis on a 1.5% agarose gel. The size of the DNA should be ~100–500 bp and is shown in Fig. 1 (see Note 3).

8. Proceed to MeDIP or store the DNA at –20°C.

3.3. MeDIP

1. Determine the concentration of the sonicated DNA.

2. Use 5 µg of DNA for each MeDIP assay. Add TE to a final volume of 450 µl in a 1.5-ml microcentrifuge tube.

3. Incubate the DNA at 95°C for 10 min and cool immediately on ice for at least 2 min.

4. Add 51 µl of 10× IP buffer.

5. Add 10 µl of 5mC antibody (thaw on ice before use, see Note 4).

6. Incubate at 4°C for 2 h on an overhead shaker.

7. Vortex the Dynabeads and pipette 30 µl for each IP.

8. Precipitate the beads with a magnetic rack (1.5–2 min). Remove the storage buffer.

9. Wash the beads two times with 800 µl PBS–BSA 0.1%.

10. Resuspend the beads in 30 µl of 1× IP buffer.

11. Add the beads to the IP mix. Incubate at 4°C for 2 h with overhead shaking.

12. Wash three times.

   (a) Precipitate beads with a magnetic rack. Remove supernatant.

   (b) Add 700 μl of 1× IP buffer.

   (c) Wash for 5 min by overhead shaking at room temperature.

13. After final wash, remove wash buffer and resuspend the beads in 250 μl digestion buffer.

14. Add 3.5 μl of proteinase K (23.4 mg/ml). Wrap the cap with parafilm.

15. Digest at 50°C for 3 h with shaking (800 rpm). Check regularly to ensure that the beads do not settle at the bottom.

16. Add 500 μl of phenol/chloroform/isoamyl alcohol. Vortex for 30 s and centrifuge at 11,000×g for 2 min. Transfer the supernatant to a new tube.

17. Repeat step 16.

18. Precipitate the DNA with 500 μl of absolute ethanol, 20 μl of 5 M NaCl, and 1 μl of glycogen (20 μg/μl).

19. Put in –20°C refrigerator overnight.

20. Centrifuge at 20,000×g for 35 min.

21. Discard the supernatant. Wash the DNA pellet with 700 μl of 70% ethanol. Centrifuge at 20,000×g for 10 min.

22. Air dry the pellet. Resuspend in 10 μl of nuclease-free water.

23. Take 2 μl for qPCR. Keep the remaining 8 μl for genome-wide amplification.

3.4. Amplification of MeDIP DNA

3.4.1. Random Priming

1. Set up the random priming reaction in a PCR tube on ice:

MeDIP DNA	8 μl
Nuclease-free water	1.15 μl
5× sequenase reaction buffer	4 μl
200 μM Primer A	4 μl
Total	17.15 μl

2. Start the random priming program of the thermocycler (see Note 5). Heat at 95°C for 4 min. Snap cool on ice and hold at 10°C.

3. Dilute 1 μl of sequenase stock (13 U/μl) with 9 μl of sequenase dilution buffer (enough for two reactions). Keep on ice.

4. Prepare the cocktail on ice (for one reaction):

10 mg/ml BSA	0.2 µl
0.1 M DTT	1 µl
10 mM dNTP mix	1.25 µl
1.3 U/µl diluted sequenase	1 µl
Total	3.45 µl

5. Add 3.45 µl of the cocktail. Mix by pipetting.

6. Put the PCR tube back to the thermocycler which is held at 10°C. Resume the program and keep the lid open.

7. Restart the program and heat again at 95°C for 4 min. Snap cool on ice and hold at 10°C.

8. Add 1 µl of the diluted sequenase. Mix by pipetting.

9. Put the PCR tube back to the thermocycler. Resume the program and keep the lid open.

10. Repeat another two rounds of the random priming (steps 7–9).

11. Add 26.4 µl TE buffer to the reaction mix (total: 50 µl).

3.4.2. PCR Amplification

1. Set up the PCR mix on ice (see Note 6):

Random primed DNA	30 µl
10× PCR buffer	20 µl
dNTP/dUTP mix	7.5 µl
100 µM Primer B	8 µl
DMSO	12 µl
50 mM MgCl$_2$	8 µl
5 U/µl Platinum Taq	4 µl
Nuclease-free water	110.5 µl
Total	200 µl

2. Place in a thermocycler and run the program:

92°C	2 min	
92°C	30 s	
40°C	30 s	
50°C	30 s	
72°C	1 min	Repeat 30 cycles
72°C	5 min	
4°C	∞	

**3.5. Cleanup
of Amplified DNA**

1. Add 24 ml of absolute ethanol to the cDNA wash buffer for the first time use of the kit.

2. In a 1.5-ml microcentrifuge tube, mix 1 ml of cDNA binding buffer with the PCR product. Vortex for 3 s.

3. Spin down briefly and load equal volume of sample to two cDNA spin columns (see Note 7).

4. Centrifuge at $8,000 \times g$ for 1 min. Discard the flow-through.

5. Transfer the cDNA spin column to a new 2-ml collection tube. Add 750 μl cDNA wash buffer to the column. Centrifuge at $8,000 \times g$ for 1 min and discard the flow-through.

6. Open the cap and centrifuge at $20,000–25,000 \times g$ for 5 min. Discard the flow-through and transfer the column to a new 1.5-ml collection tube.

7. Add 20 μl of cDNA elution buffer to the membrane of the column. Incubate at room temperature for 1 min. Then, centrifuge at $20,000–25,000 \times g$ for 1 min.

8. Repeat step 7 with another round of elution (see Note 8).

9. Pool the eluted DNA samples (from the same PCR product).

10. Take 1 μl of the sample to determine concentration.

11. Calculate the concentration.

$$\text{DNA concentration } (\mu g/\mu l) = \left(A_{260} - A_{320}\right) \times 0.05$$
$$\times \text{ dilution factor}$$

**3.6. Assessment
of Amplified Product
by qPCR**

This step is necessary before performing array hybridization as it ensures that your MeDIP product is 5mC enriched and maintained after amplification. Enrichment is shown in Fig. 2 (see Note 9).

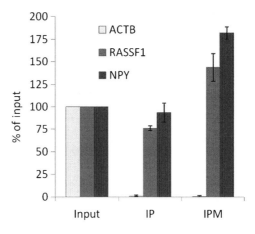

Fig. 2. Enrichment of methylated loci (RASSF1 and NPY) by MeDIP. ACTB is used as a negative-control locus.

1. The following samples are used in qPCR:

   (a) Sonicated genomic DNA (Input)

   (b) MeDIP DNA (IP)

   (c) MeDIP DNA with amplification (IPM)

2. Dilute Input DNA to 5 ng/μl, IP DNA (from step 23 of Subheading 3.3) to 40 μl, and IPM to 2 ng/μl with nuclease-free water.

3. Set up the real-time PCR master mix (one reaction):

SYBR Green Master Mix	12.5 μl
10 μM Control primer-F	1 μl
10 μM Control primer-R	1 μl
Nuclease-free water	6.5 μl
Total	21 μl

4. Make aliquots of 21 μl of master mix to each well of 96-well PCR plate.

5. Add 4 μl of Input, IP, and IPM to duplicated wells, respectively.

6. Run real-time PCR with the following condition:

95°C	10 min
95°C	15 s
60°C	1 min
95°C	1 min
Dissociation step	

7. Calculate the fold of enrichment:

   Fold of enrichment $= 2^{-(\Delta Ct\_PC - \Delta Ct\_NC)}$,
   where $\Delta Ct\_PC = Ct_{IP} - Ct_{Input}$ for positive control,
   $\Delta Ct\_NC = Ct_{IP} - Ct_{Input}$ for negative control.

3.7. Tiling Array Hybridization

3.7.1. Fragmentation of Amplified DNA

1. Set up the fragmentation mix in a PCR tube:

Double-stranded DNA	9 μg
10× cDNA fragmentation buffer	4.8 μl
10 U/μl UDG	1.5 μl
100 U/μl APE1	2.25 μl
Nuclease-free water add up to 48 μl	
Total	48 μl

2. Flick mix and spin down the tube. Incubate at 37°C for 1 h, 93°C for 2 min, and 4°C for ≥2 min.

3. (Optional) Remove 1–2 µl of sample for checking efficiency of fragmentation using a Bioanalyzer.

3.7.2. Labeling of Fragmented DNA

1. Set up the labeling mix in a PCR tube:

Fragmented DNA	45 µl
5× TdT buffer	12 µl
TdT	2 µl
5 mM DNA-labeling reagents	1 µl
Total	60 µl

2. Flick mix and spin down the tube. Incubate at 37°C for 1 h, 70°C for 10 min, and 4°C for ≥2 min.

3. (Optional) Remove 1–2 µl of sample for checking efficiency of labeling using gel-shift assay.

3.7.3. Hybridization of Labeled DNA on Tiling Arrays

1. Prepare the hybridization cocktail in a 1.5-ml microcentrifuge tube:

Fragmented and labeled DNA	60 µl
Control oligonucleotide B2	4 µl
2× hybridization mix	120 µl
DMSO	16.8 µl
Nuclease-free water	39.2 µl
Total	240 µl

2. Vortex and spin down the cocktail.

3. Heat the hybridization cocktail at 99°C for 5 min. Cool to 45°C for 5 min. Centrifuge at maximum speed for 1 min to remove any precipitate.

4. Inject 200 µl of the hybridization cocktail to the array through the lower left septum. Save the remaining cocktail at –20°C for future use.

5. Seal the septa with sticky labels to prevent leakage. Place the arrays in a 45°C hybridization oven at 60 rpm. Incubate for 16 h.

6. After hybridization, remove the hybridization cocktail and save for future use.

3.8. Array Washing, Staining, and Scanning

Refer to Affymetrix Chromatin Immunoprecipitation Assay Protocol for array washing, staining, and scanning.

4. Notes

1. In selection of kits, you should pay attention to the appropriateness of the product for your cell types (e.g., culture cells, yeast, blood cells, or fixed tissues).

2. It is critical to keep DNA on ice to avoid heat degradation.

3. DNA fragmentation is critical for efficient pull-down of methylated DNA by antibody. Prolonged sonication at low energy input usually yields smaller fragments. An average size of DNA fragments of ~200–300 bp is ideal for MeDIP.

4. Store aliquots of antibody at –20°C for long-term storage or at –4°C for up to 3 months.

5. Set up a program for random priming:

95°C	4 min
10°C	5 min, 15 s
Ramp from 10 to 37°C by holding for 23 s for each increment of 1°C.	
37°C	8 min

6. 30 µl of random-primed DNA is used for PCR amplification. If more DNA is required, scale up the PCR reaction and use all the 50 µl of random-primed DNA.

7. Each column has a binding capacity of 9–13 µg of DNA. Increase the number of columns if more DNA is synthesized from PCR.

8. If a higher concentration of DNA is desired, apply the 20 µl eluate to the same column and repeat elution.

9. At least two pairs of primers are required for assessment of specificity of MeDIP. A positive-pair control is designed to amplify regions of known methylation, whereas a negative-pair control is for regions of no methylation.

Acknowledgments

This work was supported in part by the Intramural Research Program of the National Institutes of Health (NIH), Eunice Kennedy Shriver National Institute of Child Health and Human Development, and in part by the Chinese University of Hong Kong.

References

1. Weber M, Davies JJ, Wittig D, Oakeley EJ, Haase M, Lam WL, Schubeler D (2005) Chromosome-wide and promoter-specific analyses identify sites of differential DNA methylation in normal and transformed human cells, *Nat Genet 37*, 853–862.

2. Zhang X, Yazaki J, Sundaresan A, Cokus S, Chan SW, Chen H, Henderson IR, Shinn P, Pellegrini M, Jacobsen SE, Ecker JR (2006) Genome-wide high-resolution mapping and functional analysis of DNA methylation in arabidopsis, *Cell 126*, 1189–1201.

3. Weber M, Hellmann I, Stadler MB, Ramos L, Paabo S, Rebhan M, Schubeler D (2007) Distribution, silencing potential and evolutionary impact of promoter DNA methylation in the human genome, *Nat Genet 39*, 457–466.

4. Down TA, Rakyan VK, Turner DJ, Flicek P, Li H, Kulesha E, Graf S, Johnson N, Herrero J, Tomazou EM, Thorne NP, Backdahl L, Herberth M, Howe KL, Jackson DK, Miretti MM, Marioni JC, Birney E, Hubbard TJ, Durbin R, Tavare S, Beck S (2008) A Bayesian deconvolution strategy for immunoprecipitation-based DNA methylome analysis, *Nat Biotechnol 26*, 779–785.

5. Cokus SJ, Feng S, Zhang X, Chen Z, Merriman B, Haudenschild CD, Pradhan S, Nelson S F, Pellegrini M, Jacobsen SE (2008) Shotgun bisulphite sequencing of the Arabidopsis genome reveals DNA methylation patterning, *Nature 452*, 215–219.

6. Rauch T, Li H, Wu X, Pfeifer GP (2006) MIRA-assisted microarray analysis, a new technology for the determination of DNA methylation patterns, identifies frequent methylation of homeodomain-containing genes in lung cancer cells, *Cancer Res 66*, 7939–7947.

Chapter 11

Spermatogenesis in Cryptorchidism

Alexander I. Agoulnik, Zaohua Huang, and Lydia Ferguson

Abstract

Cryptorchidism or undescended testis is the most frequent congenital abnormality in newborn boys. The process of testicular descent to the scrotum is controlled by hormones produced in Leydig cells, insulin-like3, and androgens. Variation in genetic and environmental factors might affect testicular descent. A failure of spermatogenesis and germ cell apoptosis resulting in infertility as well as an increased risk of neoplastic transformation of germ cell are the direct consequences of cryptorchidism in adulthood.

Key words: Cryptorchidism, Androgens, Insulin-like3, Spermatogenesis, Infertility, Germ cell cancer

1. Introduction

The term cryptorchidism, derived from three Greek words, "kriptós" (hidden, occult), "orchis" (testicle), and "idion" (small, diminutive), usually refers to the location of the testis outside the scrotum in the abdominal cavity, inguinal canal, or subcutaneous tissue. Cryptorchidism is one of the most common congenital defects in newborn boys. The recorded frequency of this abnormality in full-term male birth varies in different populations, but a global figure of 2–4% is generally accepted (1). In about half of all cases, the condition is spontaneously resolved within 1 year; however, if this is not the case, early surgical intervention might be beneficial.

The process of migration of the testis from its original position to the scrotum is called testicular descent. Transgenic experiments on mutant mice have led to the identification of several critical hormonal and genetic factors controlling testicular descent during embryonic development. The first, transabdominal stage of testis descent, is androgen independent and is directed by insulin-like3 (INSL3) hormone produced in testicular Leydig cells (2). The second, inguinoscrotal stage of testicular descent, is androgen dependent and any abnormality of testosterone production, androgen

Wai-Yee Chan and Le Ann Blomberg (eds.), *Germline Development: Methods and Protocols*, Methods in Molecular Biology, vol. 825, DOI 10.1007/978-1-61779-436-0_11, © Springer Science+Business Media, LLC 2012

receptor expression, or cell signaling can affect testis position (3, 4). Candidate gene association analysis of mutant alleles and cryptorchidism in human populations has indicated a possible link between the presence of gene mutations and testicular maldescent; however, the frequency of such associations is rather low and in most cases, the etiology for cryptorchidism remains unknown (2, 5).

Two major consequences of cryptorchidism are infertility and an increased risk of testicular cancer. Histological changes in cryptorchid testes include a loss of germ cells, a reduction in seminiferous tubule size, and increased fibrosis. It appears that an increase in environmental temperature of the testis is the main cause of spermatogenic arrest resulting from an inability of the undifferentiated type A_{al} spermatogonia to differentiate into the A_1 spermatogonia. Apoptosis in the cryptorchid testis affects all male germ cells with significant changes in gene expression profiles found in both somatic and germ cells; however, the causative role of individual genes in spermatogenic arrest or apoptosis remains unclear. The other consequence of cryptorchidism, an increased risk of testicular germ cell cancer, also reflects suboptimal conditions of germ cell differentiation in maldescended gonads.

2. Epidemiology of Cryptorchidism and Environmental Risk Factors

2.1. Epidemiology

The number of males affected by cryptorchidism in human populations is much debated. Trends are difficult to determine due to variation in classification criteria, study design, and follow-ups. In addition, mild cases are often undetected or underreported. In around 50% of cryptorchid cases, testes may naturally descend within the first few months after birth or are manually manipulated into the correct position. Therefore, the number of reported cases at birth is significantly higher than the prevalence of true cryptorchidism 1 year after birth (5).

Taking into account these caveats, there are still significant differences in the prevalence rates of cryptorchidism between and within populations (Table 1). A number of studies have reported a significantly higher rate in Denmark (9.0%) than in Finland (2.4%) (6). There are also reports suggesting increased rates in England (4.9%, 1984–1988), Lithuania (5.7%, 1996–1997), and the Netherlands; however, rates were not higher than those of Denmark (7). By comparison, 2–4% is generally accepted as the global average of the prevalence rate of cryptorchidism (1). In the USA, the national rate increased between the 1970s and 1980s and an increase has also been detected in South America since 1985 (8).

Cryptorchidism in England increased after the 1960s; however, recent data actually suggests a decline in rates since the 1990s (9). The study examined the prevalence of cryptorchidism in Northern

Table 1
Analysis of cryptorchidism rates in newborn boys in different countries

Country	Rate at birth (%)	Rate ≥3 months after birth (%)	Year of study	Publication
Italy	3.40	1.38	1978–1997	(112)
The UK	4.90	1.59	1984–1988	(113)
The USA	3.70	1.10	1987–1990	(114)
Northern England	0.76		1993–2000	(10)
Lithuania	5.70	1.40	1996–1997	(7)
Denmark	9.00	1.90	1997–2001	(6)
Finland	2.40	1.00	1997–1999	(6)

England between 1993 and 2000 and demonstrated a gradual decline in the number of cases. The overall prevalence was 7.6 per 1,000 male live births over the course of this 7-year period, with a rate of decrease of 0.08 cases per 1,000 male live births each year (10).

2.2. Environmental Risk Factors

The increasing prevalence of male reproductive abnormalities in some populations has led to investigations into the role environmental pollutants play in these cases. While some animal studies have indicated a direct link between chemical exposure and cryptorchidism, human studies have given less-consistent results.

The association of environmental chemicals with an increased prevalence of cryptorchidism has been characterized in a number of studies, particularly within developed countries. Man-made chemicals, such as pesticides, phthalates, bisphenol A (BPA), and polychlorinated biphenyls (PCBs), are believed to disrupt endocrine axes by mimicking estrogen interactions (11). This hypothesis was partly instigated by the finding that women treated with DES in early pregnancy gave birth to a higher rate of cryptorchid boys (12). Such disrupting chemicals have been closely linked to the suppression of INSL3 production in fetal Leydig cells, androgen function, and descent of the testes (13, 14). However, it must be kept in mind that environmental exposure to such chemicals is generally at a much lower dose than in vitro, so it is difficult to determine the contribute such disrupting chemicals may have on the prevalence of cryptorchidism.

Male reproductive health in the Western world is highly variable but shows a trend in some geographical locations. Some countries have poorer male reproductive health, for example the UK and Denmark, where the prevalence of cryptorchidism has risen (1.8% in Copenhagen 50 years ago to 8.4% 10 years ago) (6).

Contrasting the rates of cryptorchidism in Denmark with those of its close neighbor Finland suggests a contributing role of environmental factors (6). The study performed from 1997 to 2001 revealed that 9.0% of Danish boys were either unilaterally or bilaterally cryptorchid, compared to 2.4% of Finnish boys. This rate then decreased to 1.9 and 1.0%, respectively, after 3 months following either natural or manual correction (6).

Several studies have investigated the potential impact of maternal diet on cryptorchidism in their male offspring (15). One study that investigated the diet of a population of women in the Sicilian province of Ragusa detected an increased risk of cryptorchidism in the sons of women who consumed liver or smoked products more than once a week. The frequent consumption of wine was also associated with cryptorchidism. The liver, along with brain, skin, and adipose tissue, usually contains a higher concentration of PCBs than meat, which may be a contributing factor in its link to cryptorchidism (15). Another study identified a correlation between the presence of organochlorine pesticides in breast milk and the frequency of cryptorchidism, although it could not be linked to one particular compound (16). The concentration of the pesticide polybrominated diphenyl ethers (PBDEs) was significantly higher in breast milk of 62 Danish/Finnish mothers with cryptorchid sons compared to 65 control mothers.

Maternal alcohol consumption of more than five alcoholic drinks per week during pregnancy has also been linked to an increase in risk for cryptorchidism, suggesting that ethanol is able to cross the placenta and directly impact specific stages of embryo development (17). However, in other studies, no correlation was detected (18–20) indicating unreliability in the method of data collection (often maternal questionnaire) or classification methods between studies which could lead to discrepancies.

Conflicting data has been recorded in relation to maternal smoking and cryptorchidism. One study described a positive association between an increased rate of cryptorchidism and maternal smoking >10 cigarettes a day for the duration of the pregnancy (21). However, several studies did not note an association between maternal smoking and cryptorchidism (15, 20, 22).

Diabetes is the cause of many congenital malformations and the severity of which is dependent on the acuteness of the condition. A link between maternal diabetes and cryptorchidism has also been suggested, although the mechanism for the increased risk has yet to be determined (23).

Premature birth has been associated with a prevalence of cryptorchidism in a number of case studies as the inguinal stage of testicular descent occurs between weeks 26 and 35 (6, 21), although low birth weight that usually correlates with prematurity may prevent determination of the exact cause of undescended testes (21). However, low-birth-weight boys, who reach the average weight for age at 1 year, show a higher rate of spontaneous descent than those who remain below average weight indicating a role of weight in testicular descent, even after birth (7, 21).

Low birth weight, often a consequence of premature birth, has also been associated with the prevalence of cryptorchidism and has been reported in a number of case studies (24).

3. Hormonal and Genetic Factors in Testicular Descent

3.1. Two Phases of Testicular Descent

During embryonic development, the gonads differentiate from the genital ridges. After completion of sex determination, both ovary and testis remain in a high pararenal position attached to the body walls by a mesenterial ligamentous complex derived from the mesonephric mesenchyme (3, 25). At this stage, two major ligaments connect the gonads to the abdominal wall. First, the cranial mesonephric ligament that develops from the mesentery inserted between the border of the mesonephros and gonad near the hydatid region is connected to the posterior abdominal wall (25). Second, the caudal genitoinguinal ligament or the gubernaculum connects the differentiating genital ducts and later the cauda epididymis with the lower abdominal wall at the future position of the inner inguinal ring. The development and reorganization of these two ligaments, along with the differentiation of the epididymis, growth, and orientation of the gonads and reproductive tracts and finally the intraabdominal pressure, direct the movement of the testis to the scrotum.

The two-stage model of testicular descent (3) distinguishes the transabdominal phase characterized by the descent of the testis into the lower abdominal position, and the inguinoscrotal phase during which the testis moves through the inguinal canal and into the scrotum. During the transabdominal phase, the gubernacular cord and bulb are formed followed by the differentiation of muscle layers around the bulb. Further differentiation of the caudal region of the gubernaculum involves enlargement of the bulb (swelling reaction) caused by active cell proliferation and deposition of hyaluronic acid. In humans, the first stage of testicular descent occurs between 10 and 15 weeks of gestation (3). Transgenic studies in mice have identified INSL3 as the major factor in transabdominal testicular descent (26, 27). INSL3 is a small peptide hormone that belongs to the relaxin/insulin-like subfamily. It is expressed in testicular Leydig cells and is first detected right before the onset of testicular descent (28). INSL3 signals through a G protein-coupled receptor called the Relaxin Family Receptor 2 (RXFP2). Mutations of either INSL3 or RXFP2 in mice result in cryptorchid testes located in a high intraabdominal position (26, 27, 29, 30). Conversely, transgenic overexpression of INSL3 in female mice leads to gubernaculum differentiation and ovary descent to a low abdominal position (31). Combined with the fact that an androgen deficiency does not affect transabdominal descent, one can assume that INSL3 is the primary regulator of this process. Analysis of mutant gubernaculum development and the comparisons of gene expression in mutant and wild-type tissues performed in our laboratory indicated that the NOTCH and WNT/beta-catenin cell-signaling pathways might mediate the INSL3 effects at the cellular level (32). The effect of INSL3 deficiency is multifold; it causes suppression of myoblast differentiation in the muscle layers of the gubernaculum, apoptosis and reduction of androgen receptor-positive cells within the base of the gubernaculum, and a failure of processus vaginalis development.

The second inguinoscrotal stage of testicular descent is clearly androgen dependent (4). Suppression of androgen production or androgen receptor deficiency has been linked to cryptorchidism in humans and other species (33). Interestingly, an increase in testosterone production in human embryos precedes inguinoscrotal testicular descent (4). The level of serum testosterone peaks at 15–18 weeks of fetal life and declines thereafter, whereas the testis remains in the same position for 5–10 weeks after the completion of transabdominal descent. In addition, several other processes occur during this time. Starting at about 8–10 weeks, the scrotum anlage is formed and the floor of the gubernaculum base inverts to become the tunica of the sac-like processus vaginalis peritonei. The muscle layers at the rim of the gubernaculum further develop to become the wall of the cremasteric sac and the testes glide along the newly formed inguinal canal and into the scrotum (34). Because the majority of these processes are androgen sensitive, it is not surprising that many abnormalities that lead to a compromised differentiation or function of steroidogenic Leydig cells as well as the failure of androgen signaling result in various degrees of cryptorchidism (3).

A nonfunctional hypothalamo-pituitary-gonadal (HPG) axis results in hypogonadotrophic hypogonadism. The homozygous mutant mice for gonadotropin-releasing hormone, GNRH (*hpg*), GNRH receptor (*Gnrhr*), and the LH receptor knockout mouse (*LuRKO*) devoid of LH stimulation, all have impaired inguinoscrotal testicular descent (35–37). A number of human syndromes related to testicular feminization or genitourinary dysplasia have the hallmarks of androgen deficiency as well (35). Estrogen-like and antiandrogen compounds have been shown to affect testicular descent through the suppression of the functional activity of Leydig cells and decreased testosterone and INSL3 production (35). Little is known, however, about local cell-signaling pathways activated by androgen signaling in the developing gubernaculum, scrotum, cranial ligament, and epididymis. The link between inguinoscrotal cryptorchidism detected in the transgenic mouse and in some cases human mutants for several transcriptional factors, such as homeobox genes *Hoxa10* or *Hoxa11*; Wilms tumor 1, *Wt1*; ARID domain-containing protein 5B (*Arid5b*); and androgen signaling in gubernaculum development, still needs to be determined (35).

3.2. Population Genetics

Numerous studies have been conducted to determine an association of testicular maldescent in human patients with isolated cryptorchidism and the presence of mutant alleles of candidate genes. The population frequency of such mutations is usually low, population specific, and accounts only for a small portion of all analyzed cases. This is not unexpected as any mutation affecting fertility is under strong negative selection in the population. Infertility in carrier fathers prevents transmission of the mutant allele to the next generation. Thus, the propagation of the male-limited mutation can be achieved either through female transmission or through an accumulation in the population of genetic modifiers suppressing the harmful phenotype. Both of these mechanisms have been found in the case of the T222P variant allele of RXFP2 (38). Functional assays have demonstrated that the mutant receptor failed to express on the cell surface membrane. While in the Italian population, heterozygosity for T222P was strongly linked to cryptorchidism, and in all analyzed cases there was a maternal transmission of this allele; in Spain and Northern Africa, T222P heterozygotes were equally present in patient and control groups (39–41). The same appears to be true for the specific haplotype of the estrogen receptor that was reported to be associated with cryptorchidism in the Japanese population (42) or the severity of the abnormality in mixed American population (43), but not in Europe (44). Allelic variations for INSL3, HOXA10, and AR and deletions of the Y chromosome have also been linked to cryptorchidism in some, but not in all studied populations (5).

4. Spermatogenesis in Cryptorchidism

4.1. Cryptorchidism and Infertility

It is generally accepted that in most mammals, infertility in cryptorchidism is caused by a higher environmental testicular temperature. Indeed, the scrotal temperature is 2–8°C lower than the rest of the body. Experimental hyperthermia has been shown to cause similar degenerative changes as induced by cryptorchidism (45). It has been suggested that selective signaling pathways deregulated in different testicular cell populations might be responsible for spermatogenesis arrest (46). This involves a direct effect on germ cells causing cessation of spermatogonial differentiation and apoptosis as well as indirect effects, such as increased oxidative stress and abnormal energy metabolism (46–48). Changes in Sertoli cell junctions and abnormal levels of hormones produced by Leydig cells have been also associated with hyperthermia and cryptorchidism (49, 50).

The severity of infertility in human cryptorchidism depends on the position of the testes, a uni- or bilateral condition, the timing of surgical correction, and, in some cases, possible underlying endocrine or genetic pathology. It is generally accepted that an early correction within 1 year after birth might be beneficial for the preservation of the stem germ cell pool and spermatogenesis. It has been reported that newborns with intraabdominal testes have a normal number of germ cells (51). The abdominal testes are histologically normal for up to 6 months, then show a sharp decline in spermatogonia, with an empty interstitium appearing after 2 years, and a complete lack of germ cells occurring in 64% of those older than 3 years (52). By 6–8 months of age, the undescended testis shows delayed germ cell development and maturation, specifically in spermatogonial differentiation and in the appearance of primary spermatocytes (52). These changes are progressive and at puberty, germ cells become aplastic (52). At the genetic level, expression array studies of the Long Evans *orl* rat strain with inherited cryptorchidism showed relatively small differences between neonatal wild-type and cryptorchid testes, although some genes involved in muscle differentiation and cell adhesion were deregulated (53). By 11 years of age, about 35% of boys do not have germ cells in their cryptorchid testis. In unilateral cryptorchidism, the contralateral scrotal testis produces sufficient sperm for fertility and paternity is not diminished among men with a single testis compared with the general population. The success of orchidopexy (surgical correction of the testis position) in bilateral cases depends on the timing of the procedure and the position of the testes. A dramatic reduction of fertility is usually seen in patients with a history of bilateral intraabdominal cryptorchidism (52). Published data indicates significant changes in gene expression profiles in sperm from patients with a prior history of surgically corrected cryptorchidism compared to control men. Many of the identified genes have been shown to have important roles in spermatozoa metabolism, proper tail formation, and apoptosis (54).

At the histological level, cryptorchidism leads to two major consequences: arrest of spermatogonial differentiation and apoptosis of all germ cells including germ stem cells. Spermatogenesis is arrested at the undifferentiated type A spermatogonia stage. Apart from proliferation, undifferentiated spermatogonia also enter apoptosis. Spermatocytes and spermatids are most sensitive to heat-induced apoptosis. Subsequently, only undifferentiated type A spermatogonia remain in the seminiferous tubules of cryptorchid testes, with a progressive decrease in their numbers.

4.2. Apoptosis in Cryptorchidism

There are at least two broad pathways that lead to apoptosis: an "extrinsic" pathway and an "intrinsic" pathway (Fig. 1) (55, 56). In the intrinsic pathway, cellular stress factors, such as DNA damage,

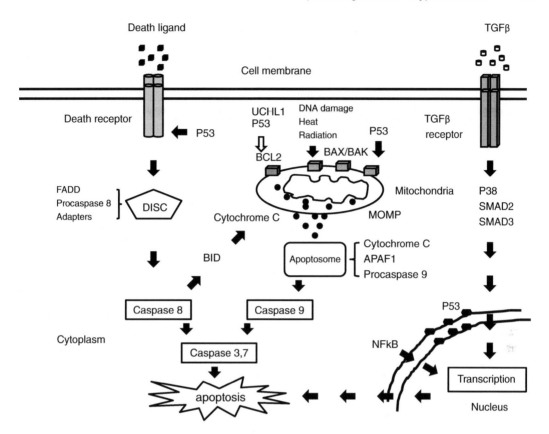

Fig. 1. Molecular mechanism of germ cell apoptosis in cryptorchidism. As somatic cells, germ cells conduct apoptosis through intrinsic and extrinsic pathways (see text for detail). P53 plays very important roles in germ cell apoptosis in cryptorchidism. It has multiple functions in the apoptosis, including inducing the expression of many proapoptotic proteins, including APAF1, BAX, BID, FAS receptor, etc., while repressing the expression of antiapoptotic proteins, like BCL2 and BCL2L1. In the cytoplasm, P53 induces the activation of BAX and BAK and increases the death receptors at the cell surface. Activated TGFβ pathway induces apoptosis by phosphorylation of SMAD2, SMAD3, and P38 and transcription modifications. UCHL1 promotes germ cell apoptosis by decreasing antiapoptotic proteins.

radiation, heat, growth factor withdrawal, and hypoxia, result in the transcriptional activation or posttranslational regulation of proapoptotic B-cell lymphoma 2 (BCL2) family proteins, BCL2-associated X protein (BAX), and BCL2 homologous antagonist/killer (BAK). The latter two factors induce permeabilization of the mitochondrial outer membrane (MOMP) and the release of mitochondrial proteins, including cytochrome C, into the cytosol. Cytochrome C associates with apoptotic protease activating factor 1 (APAF1) and recruits procaspase 9 to form the apoptosome, in which procaspase 9 is activated, leading to the apoptotic events by the activation of effector caspases 3 and 7. In the extrinsic pathway, apoptosis is trigged by the binding of death ligands, such as tumor necrosis factor (TNF) or FAS ligand, to the death receptors TNFR1 or FASR. Together with adaptor proteins, the receptor recruits procaspase 8 to form the death-inducible signaling complex (DISC). Procaspase 8 becomes active and initiates caspase 3-driven apoptosis. In some cases, caspase 8 also targets BH3-only protein, BID, resulting in the MOMP activation of the intrinsic pathway. In cellular stress, P53 is stabilized and accumulated and induces apoptosis by transcription-independent and transcription-dependent pathways (55, 56). In the nucleus, P53 can also induce the expression of many proapoptotic proteins, including APAF1, BAX, BID, and FASR while simultaneously repressing the expression of antiapoptotic proteins, such as BCL2 or BCL2L1. In the cytosol, P53 induces the activation of BAX and BAK and increases the number of death receptors at the cell surface. The question then arises as to which apoptotic pathway(s) is utilized in the germ cells of the cryptorchid testis?

By using mice with a *p53* gene-targeted allele and spontaneous mutation of *Fas*, Yin et al. (57, 58) found that these two factors play critical roles in germ cell apoptosis of cryptorchid testes. In experimentally induced cryptorchid $p53^{-/-}$ mice, germ cell apoptosis was delayed for 3 days, beginning at day 10 post surgery instead of day 7 as in the control wild-type cryptorchid group. Thus, the initial apoptotic event in cryptorchid testes is p53 dependent. An additional delay in apoptosis was then noted in mice with a double mutation of *p53* and *Fas*. Thus, FAS is responsible for the P53-independent phase of germ cell apoptosis. Since germ cells still enter apoptosis even in the double-mutant cryptochid mouse, there must be yet another pathway responsible for the programmed cell death. P53-dependent apoptosis still occurs in *Fas* mutant germ cells while FAS-dependent apoptosis happens in $p53^{-/-}$ mutant germ cells under the experimentally induced cryptorchidism. It is likely, therefore, that these two pathways are independent of each other. It is not clear why P53-dependent apoptosis precedes FAS-dependent apoptosis, although it might be related to the intensity of heat the testes are exposed to. Additionally, it has been shown that in cryptorchid monkey testes, the testicular orphan receptor 2 (TR2) is repressed by P53 via cyclin D kinase and retinoblastoma signaling (59). Li et al. (60) showed that over-expression of metastasis tumor antigen 1 (MTA1) could reduce the expression of P53 and increase spermatogenic tumor cells' resistance against heat-induced apoptosis. In the cryptorchid mouse testis, MTA1 expression decreases with the increase of P53 expression. Thus, MTA1 may act as a negative regulator of P53 to decrease cell apoptosis in the testis.

Heat-shock proteins (HSPs) have long been suspected to be involved in germ cell apoptosis in cryptorchidism. HSP105 belongs to the HSP90 family and is expressed in testicular germ cells. Analysis of the protein localization and interaction has suggested that HSP105 forms a complex with P53 at normal scrotal temperature and dissociates from it at suprascrotal temperatures in experimental cryptorchidism (61). Thus, at normal scrotal temperature, HSP105 may contribute to the cytoplasmic stabilization of P53, preventing its potential induction of apoptosis (61).

Activation of HSPs was not detected in another experiment (62); however, the activation of heat-shock transcription factor (HSF1) was discovered. Using *Hsf1* null mice, it has been determined that the deletion of HSF1 inhibits pachytene spermatocyte apoptosis in experimental cryptorchid teste while it acts as a cell-survival factor in more immature germ cells. HSP expression has been shown to be unchanged in cryptorchid testes in two other studies in mice (63) and primates (64). Another mouse mutant lacking the ubiquitin C-terminal hydrolase L1 (UCHL1) gene has also shown resistance to experimental cryptorchidism (65). UCHs belong to the deubiquitinating enzyme family, and UCHL1 expression is restricted to testicular and neural tissues. Furthermore, in cryptorchid testes of *Uchl1* null mice, the level of both antiapoptotic and prosurvival proteins has been found to be significantly higher after cryptorchid stress (65).

Ushii et al. (66) found that spermatogenic cells isolated from superoxide dismutase (SOD1) mutant mice were less resistant to heat stress compared to those isolated from wild-type control mice. Exposure to reactive oxygen species (ROS) caused apoptosis in rat spermatogenic cells in culture and superoxide scavenger application decreased the release of cytochrome C from mitochondria (66). In support of the importance of ROS, *Sod1* mutant testes with experimental cryptorchidism have an increased rate of germ cell apoptosis. Therefore, the data suggests that ROS generated in cryptorchidism causes spermatogenic cell death; ROS may also function as a signal for cell death rather than directly causing oxidative damage to cells (66).

Nitric oxide (NO), an important signaling molecule, has also been associated with apoptosis in a number of cell types. Nitric oxide synthases (NOSs) are a family of enzymes that catalyze the production of nitric oxide from L-arginine. NOSs have three isoforms, including endothelial NOS (eNOS), which is expressed in germ cells of the mouse testis. Zini et al. (67) demonstrated colocalization of eNOS protein and germ cell apoptosis in experimental cryptorchidism. These findings were further confirmed in two subsequent studies. In transgenic mice with overexpression of eNOS, Ishikawa et al. (68) discovered a significantly increased rate of germ cell apoptosis compared with a wild-type control, suggesting that NO may be involved in germ cell apoptosis. This was further confirmed in experiments with the chemical inhibition of NOS in *Hoxa11*-deficient mice with cryptorchidism; such a treatment attenuated apoptosis and improved spermatogenesis in this model (69).

Another two signal transduction pathways have been suggested to be involved in germ cell apoptosis in cryptorchid testes. In rat, Mizuno et al. (70) found that NF-kappaB signals were increased in the nucleus of apoptotic germ cells in cryptorchid testes. Yuan et al. (71) observed that the transforming growth factor beta receptor II and SMAD2 expression were also increased in the cryptorchid testis, along with an increase in the phosphorylation of SMAD2, SMAD3, and P38 and cleavage of caspase 3, indicating their involvement in apoptosis.

Finally, a potential role of cyclooxygenases (COX) in cryptorchidism-induced apoptosis was studied. COX proteins form a family of enzyme synthesizing prostanoids that includes prostaglandins, prostacyclin, and thromboxane. Pharmacological inhibition of COX can provide relief from the symptoms of inflammation and pain. Kubota et al. found that the inhibition of COX2 induced germ cell apoptosis in experimental cryptorchid testes (72). Thus, an increased COX2 expression in experimental cryptorchid mouse teste may be a protective mechanism against heat stress.

A number of other up- or downregulated genes have been identified in experimentally induced cryptorchidism in a number of rodent and primate models. Although it is difficult to ascertain the relative contribution of deregulated genes in apoptosis or germ cell differentiation, some of these genes might be indeed important. For example, a novel variant of histone H2A is a testis-specific expressed gene 1 (TSEG1) (73). TSEG1 expression increased in the surgical cryptorchid mouse in parallel with the apoptosis of spermatogenic cells. Intratesticular injection of TSEG1 also resulted in increased apoptosis of spermatogenic cells in vivo.

4.3. Somatic Cell Function in Cryptorchid Testis

In addition to germ cells, cryptorchidism also affects somatic Sertoli and Leydig cells in the testis. It has been shown that both the morphology and function of Sertoli cells, the primary supportive cells of the seminiferous epithelium, are affected in cryptorchidism. An analysis of primate Sertoli cells after induced cryptorchidism revealed significant changes in the cytoskeleton (74). Increased vimentin expression with an appearance of disorganized staining in the Sertoli cells, as well as the increased expression of cytokeratin 18, a marker of immature Sertoli cells, and the expression of desmin were observed in cryptorchid testes (74). Similar findings have also been described in rat Sertoli cells in induced cryptorchidism (75). Maekawa et al. (76) showed that the actin filament arrangement in Sertoli cells was drastically changed; the filaments became thin and disrupted compared to those of the control. Changes in cell morphology paralleled the changes in Sertoli cell gene expression. Danno et al. (77) analyzed an effect of cryptorchidism-induced heat on cold-shock protein RBM3, a member of RNA-binding motif proteins, expressed in Sertoli cells in neonatal and adult but not in fetal testis. In experimentally induced cryptorchidism, the expression of this gene was dramatically suppressed indicating temperature regulation in Sertoli cells. The FSH and androgen receptors can also be dramatically reduced in testis with induced cryptorchidism (78). While after orchidopexy the expression of these proteins was restored, the seminiferous tubules were still unable to fully support germ cell development, indicative of some irreversible changes (78).

The data on testosterone production in the cryptorchid testis is somewhat contradictory. Early reports indicated that androgen secretion in response to LH stimulation was greater in tissue from induced rat cryptorchid testes (79). In mice, however, no differences in serum testosterone levels were reported in experimentally induced cryptorchidism (80), in INSL3/RXFP2 mutant mice (26, 27, 29), or in response to a human chorionic gonadotropin (80). On the contrary, a 40–50% reduction in intratesticular androgen content, both under basal conditions and following an injection of hCG, was reported in other experiments (81). In long-term induced cryptorchidism in rats (>12 months), testosterone levels gradually decreased (82). A possible role of Sertoli cell-secreted mitogenic factor(s), affecting the functions of Leydig cell following experimental cryptorchidism, has been proposed (83). It was also suggested that cryptorchidism affects direct cell–cell interactions, essential for the germ cell supporting role of Sertoli cells and Sertoli–Leydig cell interactions. Several changes in both gap and tight junction proteins have been detected in cryptorchid testes. For example, the expression of connexin 43, an important gap junction protein localized between Leydig/Sertoli and Sertoli/germ cells, was found to be significantly decreased in cryptorchid teste (84). In the same experiments, a significant decrease of LHR, 3beta-HSD, and an increase of aromatase were also noted in Leydig cells.

4.4. Gene Expression Studies in Cryptorchidism

Recently, a number of whole-genome gene expression studies have been performed on cryptorchid testes of rodent models with induced and hereditary cryptorchidism, as well as in cryptorchid human testes (14, 45, 46, 54, 85–87). Additionally, the gene expression profiles of rodent and primate testes under hyperthermic conditions have been analyzed (45, 49). Despite an integrated pattern of gene expression detected in different somatic and germ cells in such experiments, many of the same genes and cellular pathways have been implicated as in the studies described above. The extensive analysis of deregulated genes in experimentally induced cryptorchidism performed in Dr. Yi-Xun Liu's laboratory highlighted the temporal changes in the high level of ROS related not only to the direct origin of ROS in the cryptorchid testis, but also of more upstream physiological events in energy/lipid metabolism (46). A recent study of cryptorchid testes from young boys showed a decreased or lack of expression of genes essential for the function of the hypothalamo-pituitary-testicular axis (88, 89). The authors specifically emphasized the role of early growth response protein 4 (EGR4), which is involved in regulating the secretion of LH, and fibroblast growth factor receptor 1, FGFR1, and its downstream mediators, SOS1 and RAF1, which have been previously linked to leopard and Noonan syndromes, both associated with cryptorchidism (89).

Overall, there is some progress in the dissection of the molecular mechanisms of spermatogenic arrest in cryptorchid teste in mammals. However, the overall picture remains largely elusive. The progress of targeted genetic manipulations, the analysis of new animal models, and the study of spermatogonial differentiating factors in the cryptorchid testis are needed to unmask the causes of cryptorchidism-induced infertility.

The question arises whether there is a "master" gene or biochemical pathway controlling the arrest of spermatogonial differentiation in cryptorchidism. Can germ cell differentiation arrest and apoptosis in cryptorchid testis be reversed? Rather remarkable data has been demonstrated using mice with juvenile spermatogonial depletion (*jsd*). The homozygous *jsd*$^{-/-}$ males have a single wave of spermatogenesis, followed by sterility due to a failure of type A spermatogonia to differentiate (90, 91). The *jsd* mutation is caused by a frame-shift mutation in the *Utp14b* (U3 small nucleolar ribonucleoprotein, homolog B) gene, apparently derived by retrotransposition from the X copy gene *Utp14a* (90, 91). Surprisingly, after experimentally induced cryptorchidism, spermatogenesis in the jsd mutants was, at least partially, maintained (92). Thus, a single mutation rendering an arrest of spermatogonial differentiation in a normal scrotal position was able to reverse such an arrest in an induced cryptorchid model. An investigation of mechanisms enabling the rescue of spermatogenesis awaits further analyses that may lead not only to a better understanding of germ cell differentiation, but also to new therapeutic approaches in treating infertility in cryptorchidism.

5. Cryptorchidism and Testicular Cancer

Although testicular cancer affects only 1% of males, it is the most common cancer in young men between the ages of 15–34 and, with early detection, survival rates are now around 90–95% (93, 94). A predisposition to testicular cancer in cryptorchid males was recorded as statistically significant as early as the 1940s (95). Today, an association between the two is well-documented, although somewhat conflicting concerning the strength of such an association are often disputed (96). It is now generally accepted that around 5–10% of men who develop germ cell tumors are or formerly were cryptorchid (97). The position of the cryptorchid testis has been suggested to contribute to the relative risk of testicular cancer, the greatest risk being associated with abdominally retained testes. One study that looked at cryptorchid patients between 1934 and 1975 found that 13 out of 14 cases with an abdominal testis developed a germ cell tumor (98).

Seminoma is the most common malignant tumor type associated with cryptorchidism.

Some studies have documented that males who have undergone corrective orchidopexy surgery still have an increased predisposition to testicular cancer (99) indicating a potential contribution from genetic factors in addition to positional effects. Another study estimated that the risk decreased from five- to twofold post surgery (100). In particular, prepubertal orchidopexy has been demonstrated to offer improved protection against germ cell tumors in comparison to postpubertal treatment (101). A study performed in Sweden identified 56 males with testicular cancer out of 16,983 cryptorchid men (100). Of the 56, those who underwent orchidopexy before the age of 13 had a risk of 2.23%, whereas those treated after the age of 13 had a risk of 5.40%. It is likely, therefore, that the position of the testis at puberty is a contributing factor in the risk of testicular cancer. However, this is disputed by another study that found no link between early surgery and decreased risk of germ cell tumors (102).

It has been reported that in a physically corrected testis, the risk remains higher than that of the general population, supporting hypotheses of a contributing genetic factor between cryptorchidism and testicular cancer (103, 104). However, other studies have found that the relative risk of testicular cancer in a normally descended contralateral testis opposite a cryptorchid testis is no higher than the relative risk of that of the general population (97). Biopsied testes that have been surgically moved were also found to have a higher predisposition to testicular cancer than nonbiopsied, indicating that testicular damage may contribute to the development of germ cell tumors (105).

One of the main contributing factors to an increase in germ cell tumors may be the higher temperature of the abdominal cavity in comparison to the scrotum (106, 107). Males with bilateral cryptorchid testes have an increased risk of developing germ cell tumors than unilateral cryptorchid males, and in unilateral males the cryptorchid testis is the one affected by germ cell tumors in 80% of the cases (108).

Little is known about the genetic factors or mechanisms causing the tumorigenic transformation of the germ cells in cryptorchid testis. Neither transgenic mouse models nor animals with experimentally induced cryptorchidism develop testicular cancer. Single-nucleotide polymorphism (SNP) association studies in human patients have been focused on testicular cancer patients with and without the history of cryptorchidism. The two most comprehensive studies identified a significant association of common genetic variants on two genomic regions on chromosomes 5 and 12 with testicular germ cell cancer risk. The region on chromosome 12 contained the KITLG gene, also known as stem cell factor, encoding the ligand for the receptor tyrosine kinase, c-KIT. The KITLG–KIT signaling pathway has an important role in gametogenesis, hematopoiesis, and melanogenesis (109, 110). More importantly, mutations in the *Kit* and *Kitl* genes in mice have resulted in a slightly higher incidence of germ cell tumor formation along with infertility. In a sample of 229 mice carrying a mutation in a transmembrane form of KIT, 9% were found to develop germ cell tumors, compared to 4% in a sample of 260 wild types (111). Another study identified a SNP located on chromosome 5 that links *SPRY4* (sprout-related EVH1 domain containing 2) with germ cell tumors as *SPRY4* has been shown to be regulated by the KITL–KIT pathway. However, no association between any SNPs and a history of cryptorchidism has been noted, perhaps due to the small sample size of these studies (109, 110). It is possible, therefore, that the abnormal position of the testis only confers an increased risk of tumorigenesis with additional genetic or epigenetic factors required for neoplastic germ cell transformation.

6. Future Prospective

Testicular maldescent is the most common congenital abnormality in human populations and numerous questions remain regarding the etiology and the consequences of cryptorchidism. Recent progress in understanding the genetic and hormonal control of testicular descent in animal models raises the question regarding the applicability of these results to human development. What are the most significant environmental and genetic risk factors affecting testicular descent? What are the best approaches to identify the reasons for spermatogonial arrest and germ cell apoptosis in cryptorchidism? Virtually, nothing is known about the biology and molecular mechanisms of germ cell transformation and tumorigenesis in cryptorchidism. Global gene expression studies combined with next-generation genome sequencing might lead to determining the causes of this disease. Additionally, the development of an animal model of cryptorchidism prone to germ cell tumorigenesis is needed. Further studies should focus on an understanding of these processes.

Acknowledgments

The work in Dr. Agoulnik laboratory is supported by the Eunice Kennedy Shriver National Institute of Child Health and Human Development/National Institute of Health grant R01HD37067.

References

1. Barthold JS, Gonzalez R (2003) **The epidemiology of congenital cryptorchidism, testicular ascent and orchiopexy.** *J Urol* 170:2396–2401.

2. Bogatcheva NV, Agoulnik AI (2005) **INSL3/LGR8 role in testicular descent and cryptorchidism.** *Reprod Biomed Online* 10:49–54.

3. Hutson JM, Hasthorpe S, Heyns CF (1997) **Anatomical and functional aspects of testicular descent and cryptorchidism.** *Endocr Rev* 18:259–280.

4. Hutson JM, Balic A, Nation T, Southwell B(2010) **Cryptorchidism.** *Semin Pediatr Surg* 19:215–224.

5. Foresta C, Zuccarello D, Garolla A, Ferlin A (2008) **Role of hormones, genes, and environment in human cryptorchidism.** *Endocr Rev* 29:560–580.

6. Boisen KA, Kaleva M, Main KM, Virtanen HE, Haavisto AM, Schmidt IM, Chellakooty M, Damgaard IN, Mau C, Reunanen M, Skakkebaek NE, Toppari J (2004) **Difference in prevalence of congenital cryptorchidism in infants between two Nordic countries.** *Lancet* 363:1264–1269.

7. Preiksa RT, Zilaitiene B, Matulevicius V, Skakkebaek NE, Petersen JH, Jorgensen N, Toppari J (2005) **Higher than expected prevalence of congenital cryptorchidism in Lithuania: a study of 1204 boys at birth and 1 year follow-up.** *Hum Reprod* 20:1928–1932.

8. Paulozzi LJ (1999) **International trends in rates of hypospadias and cryptorchidism.** *Environ Health Perspect* 107:297–302.

9. Jones ME, Swerdlow AJ, Griffith M, Goldacre MJ (1998) **Prenatal risk factors for cryptorchidism: a record linkage study.** *Paediatr Perinat Epidemiol* 12:383–396.

10. Abdullah NA, Pearce MS, Parker L, Wilkinson JR, Jaffray B, McNally RJ (2007) **Birth prevalence of cryptorchidism and hypospadias in northern England, 1993–2000.** *Arch Dis Child* 92:576–579.

11. Acerini CL, Hughes IA (2006) **Endocrine disrupting chemicals: a new and emerging public health problem?** *Arch Dis Child*, 91:633–641.

12. Stillman RJ (1982) **In utero exposure to diethylstilbestrol: adverse effects on the reproductive tract and reproductive performance and male and female offspring.** *Am J Obstet Gynecol*, 142:905–921.

13. Sharpe RM (2003) **The 'oestrogen hypothesis'- where do we stand now?** *Int J Androl* 26:2–15.

14. Cederroth CR, Schaad O, Descombes P, Chambon P, Vassalli JD, Nef S (2007) **Estrogen receptor alpha is a major contributor to estrogen-mediated fetal testis dysgenesis and cryptorchidism.** *Endocrinology* 148:5507–5519.

15. Giordano F, Carbone P, Nori F, Mantovani A, Taruscio D, Figa-Talamanca I (2008) **Maternal diet and the risk of hypospadias and cryptorchidism in the offspring.** *Paediatr Perinat Epidemiol* 22:249–260.

16. Main KM, Kiviranta H, Virtanen HE, Sundqvist E, Tuomisto JT, Tuomisto J, Vartiainen T, Skakkebaek NE, Toppari J (2007) **Flame retardants in placenta and breast milk and cryptorchidism in newborn boys.** *Environ Health Perspect* 115:1519–1526.

17. Damgaard IN, Jensen TK, Petersen JH, Skakkebaek NE, Toppari J, Main KM (2007) **Cryptorchidism and maternal alcohol consumption during pregnancy.** *Environ Health Perspect* 115:272–277.

18. Kurahashi N, Kasai S, Shibata T, Kakizaki H, Nonomura K, Sata F, Kishi R (2005) **Parental and neonatal risk factors for cryptorchidism.** *Med Sci Monit* 11:CR274–283.

19. Moller H, Skakkebaek NE: (1997) **Testicular cancer and cryptorchidism in relation to prenatal factors: case-control studies in Denmark.** *Cancer Causes Control* 8:904–912.

20. Biggs ML, Baer A, Critchlow CW (2002) **Maternal, delivery, and perinatal characteristics associated with cryptorchidism: a population-based case-control study among births in Washington State.** *Epidemiology*, 13:197–204.

21. Jensen MS, Toft G, Thulstrup AM, Bonde JP, Olsen J (2007) **Cryptorchidism according to maternal gestational smoking.** *Epidemiology* 18:220–225.

22. Akre O, Lipworth L, Cnattingius S, Sparen P, Ekbom A (1999) **Risk factor patterns for cryptorchidism and hypospadias.** *Epidemiology* 10:364–369.

23. Virtanen HE, Tapanainen AE, Kaleva MM, Suomi AM, Main KM, Skakkebaek NE, Toppari J (2006) **Mild gestational diabetes as a risk factor for congenital cryptorchidism.** *J Clin Endocrinol Metab* 91:4862–4865.

24. Main KM, Jensen RB, Asklund C, Hoi-Hansen CE, Skakkebaek NE (2006) **Low birth weight and male reproductive function.** *Horm Res* 65 Suppl 3:116–122.

25. Barteczko KJ, Jacob MI (2000) **The testicular descent in human. Origin, development**

and fate of the gubernaculum Hunteri, processus vaginalis peritonei, and gonadal ligaments. *Adv Anat Embryol Cell Biol* 156:III-X, 1–98.

26. Zimmermann S, Steding G, Emmen JM, Brinkmann AO, Nayernia K, Holstein AF, Engel W, Adham IM (1999) **Targeted disruption of the Insl3 gene causes bilateral cryptorchidism.** *Mol Endocrinol* 13:681–691.

27. Nef S, Parada LF (1999) **Cryptorchidism in mice mutant for Insl3.** *Nat Genet* 22:295–299.

28. Adham IM, Agoulnik AI (2004) **Insulin-like 3 signalling in testicular descent.** *Int J Androl* 27:257–265.

29. Overbeek PA, Gorlov IP, Sutherland RW, Houston JB, Harrison WR, Boettger-Tong HL, Bishop CE, Agoulnik AI (2001) **A transgenic insertion causing cryptorchidism in mice.** *Genesis* 30:26–35.

30. Gorlov IP, Kamat A, Bogatcheva NV, Jones E, Lamb DJ, Truong A, Bishop CE, McElreavey K, Agoulnik AI (2002) **Mutations of the GREAT gene cause cryptorchidism.** *Hum Mol Genet* 11:2309–2318.

31. Adham IM, Steding G, Thamm T, Bullesbach EE, Schwabe C, Paprotta I, Engel W (2002) **The overexpression of the insl3 in female mice causes descent of the ovaries.** *Mol Endocrinol* 16:244–252.

32. Kaftanovskaya EM, Feng S, Huang Z, Tan Y, Barbara AM, Kaur S, Truong A, Gorlov IP, Agoulnik AI: (2011) **Suppression of Insulin-like3 receptor reveals the role of beta-catenin and Notch signaling in gubernaculum development.** *Mol Endocrinol* 25:170–183.

33. Bay K, Main KM, Toppari J, Skakkebaek NE: (2011) **Testicular descent: INSL3, testosterone, genes and the intrauterine milieu.** *Nat Rev Urol* March 15 [Epub ahead of print] PMID 21403659.

34. van der Schoot P (1996) **Towards a rational terminology in the study of the gubernaculum testis: arguments in support of the notion that the cremasteric sac should be considered the gubernaculum in postnatal rats and other mammals.** *J Anat* 189 (Pt 1):97–108.

35. Klonisch T, Fowler PA, Hombach-Klonisch S (2004) **Molecular and genetic regulation of testis descent and external genitalia development.** *Dev Biol* 270:1–18.

36. Feng S, Bogatcheva NV, Truong A, Engel W, Adham IM, Agoulnik AI (2006) **Over expression of insulin-like 3 does not prevent cryptorchidism in GNRHR or HOXA10 deficient mice.** *J Urol* 176:399–404.

37. Pask AJ, Kanasaki H, Kaiser UB, Conn PM, Janovick JA, Stockton DW, Hess DL, Justice MJ, Behringer RR (2005) **A novel mouse model of hypogonadotrophic hypogonadism: N-ethyl-N-nitrosourea-induced gonadotropin-releasing hormone receptor gene mutation.** *Mol Endocrinol* 19:972–981.

38. Bogatcheva NV, Ferlin A, Feng S, Truong A, Gianesello L, Foresta C, Agoulnik AI (2007) **T222P mutation of the insulin-like 3 hormone receptor LGR8 is associated with testicular maldescent and hinders receptor expression on the cell surface membrane.** *Am J Physiol Endocrinol Metab* 292:E138–144.

39. El Houate B, Rouba H, Imken L, Sibai H, Chafik A, Boulouiz R, Chadli E, Hassar M, McElreavey K, Barakat A (2008) **No association between T222P/LGR8 mutation and cryptorchidism in the Moroccan population.** *Horm Res* 70:236–239.

40. Feng S, Ferlin A, Truong A, Bathgate R, Wade JD, Corbett S, Han S, Tannour-Louet M, Lamb DJ, Foresta C, Agoulnik AI (2009) **INSL3/RXFP2 signaling in testicular descent.** *Ann N Y Acad Sci* 1160:197–204.

41. Ars E, Lo Giacco D, Bassas L, Nuti F, Rajmil O, Ruiz P, Garat JM, Ruiz-Castane E, Krausz C (2010) **Further insights into the role of T222P variant of RXFP2 in non-syndromic cryptorchidism in two Mediterranean populations.** *Int J Androl* July 16 [Epub ahead of print] PMID20636340.

42. Yoshida R, Fukami M, Sasagawa I, Hasegawa T, Kamatani N, Ogata T (2005) **Association of cryptorchidism with a specific haplotype of the estrogen receptor alpha gene: implication for the susceptibility to estrogenic environmental endocrine disruptors.** *J Clin Endocrinol Metab* 90:4716–4721.

43. Wang Y, Barthold J, Figueroa E, Gonzalez R, Noh PH, Wang M, Manson J (2008) **Analysis of five single nucleotide polymorphisms in the ESR1 gene in cryptorchidism.** *Birth Defects Res A Clin Mol Teratol* 82:482–485.

44. Galan JJ, Guarducci E, Nuti F, Gonzalez A, Ruiz M, Ruiz A, Krausz C (2007) **Molecular analysis of estrogen receptor alpha gene AGATA haplotype and SNP12 in European populations: potential protective effect for cryptorchidism and lack of association with male infertility.** *Hum Reprod* 22:444–449.

45. Li Y, Zhou Q, Hively R, Yang L, Small C, Griswold MD (2009) **Differential gene expression in the testes of different murine strains under normal and hyperthermic conditions.** *J Androl* 30:325–337.

46. Li YC, Hu XQ, Xiao LJ, Hu ZY, Guo J, Zhang KY, Song XX, Liu YX (2006)

An oligonucleotide microarray study on gene expression profile in mouse testis of experimental cryptorchidism. *Front Biosci* 11:2465–2482.

47. Ahotupa M, Huhtaniemi I (1992) **Impaired detoxification of reactive oxygen and consequent oxidative stress in experimentally cryptorchid rat testis.** *Biol Reprod* **46:** 1114–1118.

48. Peltola V, Huhtaniemi I, Ahotupa M (1995) **Abdominal position of the rat testis is associated with high level of lipid peroxidation.** *Biol Reprod* 53:1146–1150.

49. Liu Y, Li X (2010) **Molecular basis of cryptorchidism-induced infertility.** *Sci China Life Sci* 53:1274–1283.

50. Lee PA, Coughlin MT (2002) **Leydig cell function after cryptorchidism: evidence of the beneficial result of early surgery.** *J Urol* 167:1824–1827.

51. Hadziselimovic F, Thommen L, Girard J, Herzog B (1986) **The significance of postnatal gonadotropin surge for testicular development in normal and cryptorchid testes.** *J Urol* 136:274–276.

52. Taran I, Elder JS (2006) **Results of orchiopexy for the undescended testis.** *World J Urol* 24:231–239.

53. Barthold JS, McCahan SM, Singh AV, Knudsen TB, Si X, Campion L, Akins RE (2008) **Altered expression of muscle- and cytoskeleton-related genes in a rat strain with inherited cryptorchidism.** *J Androl* 29:352–366.

54. Nguyen MT, Delaney DP, Kolon TF (2009) **Gene expression alterations in cryptorchid males using spermatozoal microarray analysis.** *Fertil Steril* 92:182–187.

55. Chipuk JE, Green DR (2006) **Dissecting p53-dependent apoptosis.** *Cell Death Differ* 13:994–1002.

56. Speidel D (2010) **Transcription-independent p53 apoptosis: an alternative route to death.** *Trends Cell Biol* 20:14–24.

57. Yin Y, DeWolf WC, Morgentaler A (1998) **Experimental cryptorchidism induces testicular germ cell apoptosis by p53-dependent and -independent pathways in mice.** *Biol Reprod* 58:492–496.

58. Yin Y, Stahl BC, DeWolf WC, Morgentaler A (2002) **P53 and Fas are sequential mechanisms of testicular germ cell apoptosis.** *J Androl* 23:64–70.

59. Mu X, Liu Y, Collins LL, Kim E, Chang C (2000) **The p53/retinoblastoma-mediated repression of testicular orphan receptor-2 in the rhesus monkey with cryptorchidism.** *J Biol Chem* 275:23877–23883.

60. Li W, Bao W, Ma J, Liu X, Xu R, Wang RA, Zhang Y (2008) **Metastasis tumor antigen 1 is involved in the resistance to heat stress-induced testicular apoptosis.** *FEBS Lett* 582:869–873.

61. Kumagai J, Fukuda J, Kodama H, Murata M, Kawamura K, Itoh H, Tanaka T (2000) **Germ cell-specific heat shock protein 105 binds to p53 in a temperature-sensitive manner in rat testis.** *Eur J Biochem* 267:3073–3078.

62. Izu H, Inouye S, Fujimoto M, Shiraishi K, Naito K, Nakai A (2004) **Heat shock transcription factor 1 is involved in quality-control mechanisms in male germ cells.** *Biol Reprod* 70:18–24.

63. Widlak W, Winiarski B, Krawczyk A, Vydra N, Malusecka E, Krawczyk Z (2007) **Inducible 70 kDa heat shock protein does not protect spermatogenic cells from damage induced by cryptorchidism.** *Int J Androl* 30:80–87.

64. Zhou XC, Han XB, Hu ZY, Zhou RJ, Liu YX (2001) **Expression of Hsp70-2 in unilateral cryptorchid testis of rhesus monkey during germ cell apoptosis.** *Endocrine* 16:89–95.

65. Kwon J, Wang YL, Setsuie R, Sekiguchi S, Sato Y, Sakurai M, Noda M, Aoki S, Yoshikawa Y, Wada K (2004) **Two closely related ubiquitin C-terminal hydrolase isozymes function as reciprocal modulators of germ cell apoptosis in cryptorchid testis.** *Am J Pathol* 165:1367–1374.

66. Ishii T, Matsuki S, Iuchi Y, Okada F, Toyosaki S, Tomita Y, Ikeda Y, Fujii J (2005) **Accelerated impairment of spermatogenic cells in SOD1-knockout mice under heat stress.** *Free Radic Res* 39:697–705.

67. Zini A, Schlegel PN (1997) **Cu/Zn superoxide dismutase, catalase and glutathione peroxidase mRNA expression in the rat testis after surgical cryptorchidism and efferent duct ligation.** *J Urol* 158:659–663.

68. Ishikawa T, Kondo Y, Goda K, Fujisawa M (2005) **Overexpression of endothelial nitric oxide synthase in transgenic mice accelerates testicular germ cell apoptosis induced by experimental cryptorchidism.** *J Androl* 26:281–288.

69. DeFoor WR, Kuan CY, Pinkerton M, Sheldon CA, Lewis AG (2004) **Modulation of germ cell apoptosis with a nitric oxide synthase inhibitor in a murine model of congenital cryptorchidism.** *J Urol* 172:1731–1735; discussion 1735.

70. Mizuno K, Hayashi Y, Kojima Y, Nakane A, Tozawa K, Kohri K (2009) **Activation of NF-kappaB associated with germ cell apoptosis in testes of experimentally induced cryptorchid rat model.** *Urology* 73:389–393.

71. Yuan JL, Zhang YT, Wang Y (2010) **Increased apoptosis of spermatogenic cells in cryptorchidism rat model and its correlation with transforming growth factor beta type II receptor.** *Urology* 75:992–998.

72. Kubota H, Sasaki S, Kubota Y, Umemoto Y, Yanai Y, Tozawa K, Hayashi Y, Kohri K (2011) **Cyclooxygenase-2 Protects Germ Cells Against Spermatogenesis Disturbance in Experimental Cryptorchidism Model Mice.** *J Androl* 32(1):77–85.

73. Gu C, Tong Q, Zheng L, Liang Z, Pu J, Mei H, Hu T, Du Z, Tian F, Zeng F (2010) **TSEG-1, a novel member of histone H2A variants, participates in spermatogenesis via promoting apoptosis of spermatogenic cells.** *Genomics* 95:278–289.

74. Zhang ZH, Hu ZY, Song XX, Xiao LJ, Zou RJ, Han CS, Liu YX (2004) **Disrupted expression of intermediate filaments in the testis of rhesus monkey after experimental cryptorchidism.** *Int J Androl* 27:234–239.

75. Wang ZQ, Watanabe Y, Toki A, Itano T (2002) **Altered distribution of Sertoli cell vimentin and increased apoptosis in cryptorchid rats.** *J Pediatr Surg* 37:648–652.

76. Maekawa M, Kazama H, Kamimura K, Nagano T (1995) **Changes in the arrangement of actin filaments in myoid cells and Sertoli cells of rat testes during postnatal development and after experimental cryptorchidism.** *Anat Rec* 241:59–69.

77. Danno S, Itoh K, Matsuda T, Fujita J (2000) **Decreased expression of mouse Rbm3, a cold-shock protein, in Sertoli cells of cryptorchid testis.** *Am J Pathol* 156:1685–1692.

78. Monet-Kuntz C, Barenton B, Locatelli A, Fontaine I, Perreau C, Hochereau-de Reviers MT (1987) **Effects of experimental cryptorchidism and subsequent orchidopexy on seminiferous tubule functions in the lamb.** *J Androl* 8:148–154.

79. Jansz GF, Pomerantz DK (1986) **A comparison of Leydig cell function after unilateral and bilateral cryptorchidism and efferent-duct-ligation.** *Biol Reprod* 34:316–321.

80. Mendis-Handagama SM, Kerr JB, de Kretser DM (1990) **Experimental cryptorchidism in the adult mouse: II. A hormonal study.** *J Androl* 11:548–554.

81. Murphy L, O'Shaughnessy PJ (1991) **Effect of cryptorchidism on testicular and Leydig cell androgen production in the mouse.** *Int J Androl* 14:66–74.

82. Hedger MP, McFarlane JR, de Kretser DM, Risbridger GP (1994) **Multiple factors with steroidogenesis-regulating activity in testicular intertubular fluid from normal and** experimentally cryptorchid adult rats. *Steroids* 59:676–685.

83. Wu N, Murono EP: (1996) **Temperature and germ cell regulation of Leydig cell proliferation stimulated by Sertoli cell-secreted mitogenic factor: a possible role in cryptorchidism.** *Andrologia* 28:247–257.

84. Hejmej A, Kotula-Balak M, Sadowska J, Bilinska (2007) **Expression of connexin 43 protein in testes, epididymides and prostates of stallions.** *Equine Vet J* 39:122–127.

85. Mizuno K, Kojima Y, Kurokawa S, Maruyama T, Sasaki S, Kohri K, Hayashi Y (2009) **Identification of differentially expressed genes in human cryptorchid testes using suppression subtractive hybridization.** *J Urol* 181:1330–1337; discussion 1337.

86. Hirai T, Tsujimura A, Ueda T, Fujita K, Matsuoka Y, Takao T, Miyagawa Y, Koike N, Okuyama A: (2009) **Effect of 1,25-dihydroxyvitamin d on testicular morphology and gene expression in experimental cryptorchid mouse: testis specific cDNA microarray analysis and potential implication in male infertility.** *J Urol* 181:1487–1492.

87. Orwig KE, Ryu BY, Master SR, Phillips BT, Mack M, Avarbock MR, Chodosh L, Brinster RL (2008) **Genes involved in post-transcriptional regulation are overrepresented in stem/progenitor spermatogonia of cryptorchid mouse testes.** *Stem Cells* 26:927–938.

88. Hadziselimovic F, Hadziselimovic NO, Demougin P, Krey G, Hoecht B, Oakeley EJ (2009) **EGR4 is a master gene responsible for fertility in cryptorchidism.** *Sex Dev* 3:253–263.

89. Hadziselimovic NO, de Geyter C, Demougin P, Oakeley EJ, Hadziselimovic F (2010) **Decreased expression of FGFR1, SOS1, RAF1 genes in cryptorchidism.** *Urol Int* 84:353–361.

90. Rohozinski J, Bishop CE: (2004) **The mouse juvenile spermatogonial depletion (jsd) phenotype is due to a mutation in the X-derived retrogene, mUtp14b.** *Proc Natl Acad Sci USA* 101:11695–11700.

91. Bradley J, Baltus A, Skaletsky H, Royce-Tolland M, Dewar K, Page DC (2004) **An X-to-autosome retrogene is required for spermatogenesis in mice.** *Nat Genet* 36:872–876.

92. Shetty G, Weng CC (2004) **Cryptorchidism rescues spermatogonial differentiation in juvenile spermatogonial depletion (jsd) mice.** *Endocrinology* 145:126–133.

93. Ries L, Melbert, D., Krapcho, M et al (2008) **National Cancer Institute 1975–2005.** *SEER cancer statistics review.*

94. Kinkade S (1999) **Testicular cancer.** *Am Fam Physician*, **59**:2539–2544, 2549–2550.

95. Campbell H (1942) **Incidence of malignant growth of the undescended testicle.** *Arch Surg* **44**:353–369.

96. Wood HM, Elder JS (2009) **Cryptorchidism and testicular cancer: separating fact from fiction.** *J Urol* **181**:452–461.

97. Kanto S, Hiramatsu M, Suzuki K, Ishidoya S, Saito H, Yamada S, Satoh M, Saito S, Fukuzaki A, Arai Y (2004) **Risk factors in past histories and familial episodes related to development of testicular germ cell tumor.** *Int J Urol* **11**:640–646.

98. Batata MA, Whitmore WF, Jr., Chu FC, Hilaris BS, Loh J, Grabstald H, Golbey R (1980) **Cryptorchidism and testicular cancer.** *J Urol* **124**:382–387.

99. Dieckmann KP, Pichlmeier U: (2004) **Clinical epidemiology of testicular germ cell tumors.** *World J Urol* **22**:2–14.

100. Pettersson A, Richiardi L, Nordenskjold A, Kaijser M, Akre O: (2007) **Age at surgery for undescended testis and risk of testicular cancer.** *N Engl J Med* **356**:1835–1841.

101. Walsh TJ, Dall'Era MA, Croughan MS, Carroll PR, Turek PJ (2007) **Prepubertal orchiopexy for cryptorchidism may be associated with lower risk of testicular cancer.** *J Urol* **178**:1440–1446; discussion 1446.

102. Myrup C, Schnack TH, Wohlfahrt J (2007) **Correction of cryptorchidism and testicular cancer.** *N Engl J Med* **357**:825–827; author reply 825–827.

103. Mathers MJ, Sperling H, Rubben H, Roth S (2009) **The undescended testis: diagnosis, treatment and long-term consequences.** *Dtsch Arztebl Int* **106**:527–532.

104. Akre O, Richiardi L (2009) **Does a testicular dysgenesis syndrome exist?** *Hum Reprod* **24**:2053–2060.

105. Swerdlow AJ, Higgins CD, Pike MC: (1997) **Risk of testicular cancer in cohort of boys with cryptorchidism.** *BMJ* **314**:1507–1511.

106. Mieusset R, Fouda PJ, Vaysse P, Guitard J, Moscovici J, Juskiewenski S (1993) **Increase in testicular temperature in case of cryptorchidism in boys.** *Fertil Steril* **59**:1319–1321.

107. Guminska A, Slowikowska-Hilczer J, Kuzanski W, Sosnowski M, Oszukowska E, Marchlewska K, Walczak-Jedrzejowska R, Niedzielski J, Kula K (2007) **Features of impaired seminiferous tubule differentiation are associated with germ cell neoplasia in adult men surgically treated in childhood because of cryptorchidism.** *Folia Histochem Cytobiol* **45** Suppl 1:S163-168.

108. Martin DC (1982) **Malignancy in the cryptorchid testis.** *Urol Clin North Am* **9**:371–376.

109. Kanetsky PA, Mitra N, Vardhanabhuti S, Li M, Vaughn DJ, Letrero R, Ciosek SL, Doody DR, Smith LM, Weaver J, Albano A, Chen C, Starr JR, Rader DJ, Godwin AK, Reilly MP, Hakonarson H, Schwartz SM, Nathanson KL (2009) **Common variation in KITLG and at 5q31.3 predisposes to testicular germ cell cancer.** *Nat Genet* **41**:811–815.

110. Rapley EA, Turnbull C, Al Olama AA, Dermitzakis ET, Linger R, Huddart RA, Renwick A, Hughes D, Hines S, Seal S, Morrison J, Nsengimana J, Deloukas P, UK Testicular Cancer Collaboration, Rahman N, Bishop DT, Easton DF, Stratton MR (2009) **A genome-wide association study of testicular germ cell tumor.** *Nat Genet* **41**:807–810.

111. Heaney JD, Lam MY, Michelson MV, Nadeau JH: (2008) **Loss of the transmembrane but not the soluble kit ligand isoform increases testicular germ cell tumor susceptibility in mice.** *Cancer Res* **68**:5193–5197.

112. Ghirri P, Ciulli C, Vuerich M, Cuttano A, Faraoni M, Guerrini L, Spinelli C, Tognetti S, Boldrini A (2002) **Incidence at birth and natural history of cryptorchidism: a study of 10,730 consecutive male infants.** *J Endocrinol Invest* **25**:709–715.

113. Baldessarini RJ, Kula NS, Campbell A, Bakthavachalam V, Yuan J, Neumeyer JL (1992) **Prolonged D2 antidopaminergic activity of alkylating and nonalkylating derivatives of spiperone in rat brain.** *Mol Pharmacol* **42**:856–863.

114. Berkowitz GS, Lapinski RH, Dolgin SE, Gazella JG, Bodian CA, Holzman IR (1993) **Prevalence and natural history of cryptorchidism.** *Pediatrics* **92**:44–49.

Part II

Female Germline

Chapter 12

In Vitro Culture of Fetal Ovaries: A Model to Study Factors Regulating Early Follicular Development

Shyamal K. Roy, Cheng Wang, Anindit Mukherjee, and Prabuddha Chakraborty

Abstract

Follicular development commences with the formation of primordial follicles, which begins with the differentiation of pluripotent ovarian somatic cells into early granulosa cells and their apposition to the oocytes in the egg nest. The process of primordial follicle morphogenesis and factors affecting the formation and development of primordial follicles can be examined in vitro using fetal ovaries in organ culture. The functions of candidate genes involved in primordial folliculogenesis can be examined using siRNA or shRNA, which can knockdown specific mRNA targets at specific time points. Here, we describe the organ culture protocol for fetal hamster ovary with GPR30 siRNA as an example. The method to morphologically analyze follicular development is also discussed.

Key words: Primordial follicle, In vitro, Oocytes, siRNA, Method

1. Introduction

Normal follicular development in mammalian ovaries is essential for successful pregnancies. Ovarian follicles harbor the oocyte, which carries the genetic blueprint of the female and is eventually expelled from the follicle during ovulation (1–4). Follicular development commences with the formation of primordial follicles, which are the very early stage oocytes surrounded by one or few flattened early granulosa cells (1). Morphogenesis of primordial follicles begins with the differentiation of pluripotent ovarian somatic cells into early granulosa cells and their apposition to the oocytes in the egg nest (1, 5–8). Throughout the reproductive life span, primordial follicles transition into primary follicles, which have a single layer of cuboidal granulosa cells (GC) surrounded by

Wai-Yee Chan and Le Ann Blomberg (eds.), *Germline Development: Methods and Protocols*, Methods in Molecular Biology, vol. 825, DOI 10.1007/978-1-61779-436-0_12, © Springer Science+Business Media, LLC 2012

a basement membrane, but no distinct thecal cells, as they enter in the growth pool (2). Because the primordial follicle pool represents the entire repertoire of the follicular reserve, abnormalities in the formation or development of primordial follicles is expected to impact fertility and fecundity. A loss of function, such as a mutation in FSH receptors in human leads to impairment of primordial follicle formation with consequent infertility (9).

Factors affecting the formation and development of primordial follicles can be examined in vitro using fetal ovaries in organ culture. If the goal is to determine the mechanisms underlying follicular morphogenesis, the functions of candidate genes can be examined using small interfering RNA (siRNA) or small hairpin RNA (shRNA) to knockdown specific targets at specific time points. This approach allows conditional, and tissue-specific knockdown relatively easily and quickly. The approach is dependent on the identification of an effective unique siRNA and an efficient mode of delivery to the cells of a multicellular organ. However, recent advancement in viral delivery methods using adenovirus or lentivirus vector holds promise.

When studying primordial follicle formation in vitro, it is important that ample time is available for a molecular reagent to interfere with the morphogenetic process. The hamster model is remarkably suited for this purpose. The hamster has a 16-day gestational period with parturition occurring the afternoon of 16th day. However, morphologically distinct primordial follicles cannot be identified until postnatal day 8 (P8). This offers a unique, long window to examine factors modulating follicular morphogenesis. One limitation of the hamster model is that a reference genome database does not exist for this species. However, many transcript sequences cloned so far have significant similarity with those of the mouse. Therefore, notwithstanding this limitation, we have successfully used the siRNA approach to establish the importance of various factors in the process of primordial folliculogenesis.

The principle and mechanism of siRNA knockdown of target genes have been discussed in many recent reviews (10–12). The siRNA approach relies on the targeting of specific mRNA in cells using synthetic RNA 21–22 nt in length, which acts as a guide for the degradation of specific mRNA (13–15). Because of the need for relatively high target specificity, the siRNA must be species-specific or at least the siRNA sequence must be 100% similar to the target mRNA sequence. Despite the development of guidelines to design highly effective siRNA sequences for mammalian RNA interference (16) many recent studies have indicated considerable off-target effects of siRNA (15, 17). Because the end result of an siRNA knockdown is often changes in cell functions, the specificity of the siRNA is critical for accurate interpretation of the results. For example, the end result of siRNA knockdown of a cell survival gene is expected to be cell death; however, cell death can occur indirectly if the siRNA adversely targets gene transcripts that are critical for cell metabolism, but unrelated to the gene of interest. Therefore, end results do not necessarily indicate a specific action of siRNA. Fortunately, Web-based online software offered by siRNA manufacturing companies, such as ThermoFisher-Dharmacon and Ambion/Invitrogen allows investigators to design siRNA with greater specificity. An extended service is also available from the manufacturers for a fee. A Web-based online freeware (http://www.design.RNA.jp) promises designing highly effective target-specific siRNA sequences (15) with negligible off-target effect. Additionally, pretested siRNA for a plethora of human-, mouse-, and rat-specific genes are readily available from many companies, such as Dharmacon/Thermo-Fisher, Ambion/Invitrogen, etc to name a few. These tested reagents can eliminate time-consuming design and synthesis exercises. However, for species without a genome database, such as the golden hamster, specific siRNA must be designed using indicated software and comparison of nucleic acid sequence with closely related species, such as the mouse (18, 19).

The second major challenge is efficient delivery of siRNA to cells. The degree of difficulty increases many fold for organ culture where siRNA must reach cells located at different depths. Many lipid-based transfection agents are available, but in our hands, Metafectane (Biontex) works best (18). Lentivirus-mediated shRNA strategy can overcome some of the limitations. The chapter describes in detail the methodology for the culture of the fetal hamster ovary, in vitro transfection with siRNA, identification of primordial follicles and data analysis. We routinely use the described siRNA technique to investigate the role of estrogen in primordial folliculogenesis via the G-protein coupled receptor 30 (GPR30) by siRNA knockdown of GPR30 The staging of follicles during their growth phase is also described.

2. Materials

All ovary collection and culture media should be prepared in autoclaved 20.8 Ω MilliQ water and filtered through a sterile 0.2 mm cellulose acetate filter from Millipore (Miilipore, Billerica, MA). We use media from GIBCO/Invitrogen (Invitrogen, Carlsbad, CA). Because the protocol is for organ culture, all necessary precautions to maintain utmost sterility must be strictly followed (see Note 1). This will prevent fungal contamination of the cultures.

2.1. Ovary Collection

1. A stereozoom dissecting scope (Leica) to be used for dissection and ovary collection. This scope should be in the laboratory in an aseptic place other than the culture room. It is important to have a wider platform underneath the zoom head so that the dissection can be performed. The advantage of a zoom microscope is to change the magnification incrementally for viewing convenience.

2. A second stereozoom microscope (Leica) with a digital camera to be used for cleaning and handling isolated ovaries in the laminar flow hood. This scope should not be used for necropsy.

3. Two pairs of jeweler's forceps (Fine Science Tools, Inc, Dumoxel Biologue, Foster City, CA), one for dissecting ovaries from the fetuses and one for cleaning ovaries in the hood. Both should be cleaned with soap and warm water, rinsed with 70% ethanol, and wiped dry after each use.

4. Two pairs of spring-handle fine-point iris scissors, one for dissecting ovaries from the fetuses and one for cleaning ovaries in the hood. Both should be cleaned with soap and warm water, rinsed with 70% ethanol, and wiped dry after each use (see Note 2).

5. A pair of dissecting scissors, and a pair of small fine-tip forceps.

6. 70% ethanol: dilute 70 ml of 95% ethyl alcohol to 95 ml with MilliQ water. Place in a squirt bottle near the dissecting area and one in the hood. Ethanol will sterilize the water and bottle inside, and will remain sterile.

7. Phosphate-buffered saline, pH 7.4: autoclaved and placed in a squirt bottle at room temperature.

8. Clean paper towels placed near the dissection scope.

9. Sterile 60-mm in vitro fertilization culture dishes with cover (Fig. 1a). This can be substituted with 35-mm petri dishes for economy.

10. A set of Eppendorf micropipettors ranging from 1 to 1,000 ml capacity.

11. Sterile 1.5- and 0.65-ml polypropylene microtubes. We use Avant microtubes supplied by Midwest Scientific (St. Louis, MO) because these are very easy to open and close, glass-clear

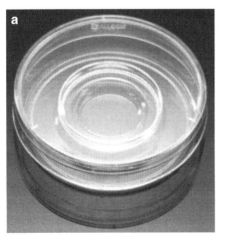
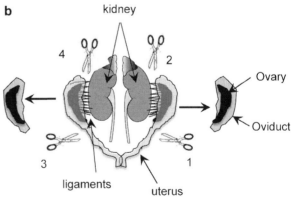

Fig. 1. (**a**) Image of an IVF culture dish (**a**), and (**b**) diagrammatic presentation of ovary–oviduct complex with relation to the kidney. The structure on both sides represents the cut ovary–oviduct complex.

with the molding point on its side that allows an unobstructed view of small pellet.

12. Sodium pentobarbital (Nembutal) for euthanasia: 15 mg/kg body weight in sterile saline made fresh before use.

13. A Bell jar connected with a CO_2 tank with gas regulator for euthanasia (see Note 3).

14. Gloves and mask.

15. 15-day pregnant hamsters (see Note 4).

16. Autoclaved borosilicate Pasteur pipettes: 6 inch.

2.2. Ovary Culture

1. Twelve-well (TRP) or 24-well (Falcon) sterile tissue culture cluster plate, and 12-well (Falcon) or 24-well (Falcon) inserts with 3 μmm pore size mesh for cluster plate. These inserts are nontissue-culture treated, which prevent ovarian cells to spread on the mesh and migrate through the pore. This is important for detaching the ovaries from the mesh after the culture.

2. Factors to be tested; we use, for example estradiol17β (E2).

3. siRNA for appropriate target and suitable delivery agent. Our work, for examples uses GPR30 and Metafectane as examples. For initial optimization use a FITC-labeled nontargeting siRNA. This nontargeting siRNA is a 19-nucleotide long single stranded nucleic acid, which is checked for no probable matching in human, mouse or rat genome database.

2.2.1. Base Culture Medium (A)

(a) Dulbecco's modified Eagle's medium (DMEM) without phenol red, glutamine, Na-pyruvate and with 1,000 mg/L glucose.

(b) 200 mM GlutaMAX.

(c) 100 mM Na-pyruvate.

(d) Antibiotic–antimycotic: 10,000 U penicillin G, 10,000 μg streptomycin, and 25 μg amphotericin B per ml.

(e) Bovine serum albumin.

Store solutions b through d frozen at –20°C in 5 ml aliquots (see Note 5); see Subheading 3. The base medium can be changed to suit the experimental need. Hamster ovaries grow well in DMEM with low glucose (1,000 mg/L) (see Note 6).

2.2.2. Complete Culture Medium (B)

(a) Culture medium A

(b) Human holo-transferrin (BD Bioscience) 10 mg lyophilized: dissolve the powder in 1 ml autoclaved MilliQ water and store 0.1 ml aliquots in sterile 0.65-ml PP tubes at –80°C.

(c) Human recombinant insulin (Sigma/Aldrich): ready to use solution: Store aliquots at –80°C, and avoid freeze–thaw.

(d) Linoleic acid/albumin complex (BD Biosciences): 2.5/500 mg-lyophilized: three vials are needed. Store at 4°C and resuspended each vial with 0.5 ml autoclaved MilliQ water before use. Store any excess at 4°C. Do not freeze.

(e) Selenous acid: (BD Biosciences) 100 mg powder: make 1 mg/0.1 ml stock I by dissolving the entire vial content in 10 ml autoclaved MilliQ water. Aliquot 0.5 ml in 0.65 ml PP tubes. Make a stock II by diluting 0.1 ml stock I to 1 ml with MilliQ water; the concentration is 1 μg/μl. Aliquot 0.05 ml in 0.65-ml PP tubes. Store all stocks at –80°C. Stock II can be refrozen 2–3 times.

(f) Bovine serum albumin (BSA): make a 30–40% solution of BSA (A33110; see Note 7) in autoclaved MilliQ water.

2.3. Post-culture Ovary Processing

(a) Bouin's Fixative: 75 ml saturated solution of picric acid in MilliQ water, 25 ml deionized (filtered over a Dowex 50 resin, Bio-Rad, CA) formaldehyde and 5 ml glacial acetic acid. Prepare this fixative in advance and store at room temperature. This solution is stable for months in a closed bottle.

(b) Lithium carbonate solution.

(c) JB-4.

(d) DPX.

(e) An upright or inverted bright-field research microscope. We use a Leica DM55 research microscope with PlanApo objectives.

(f) Digital high-resolution color camera. We use a Leica DFC320 camera.

(g) Image acquisition and analysis software. We use the Openlab application for image capture and the Image pro Plus application for analysis.

3. Methods

3.1. Preparation of the Base Culture Medium (A)

1. Thaw a 5-ml tube each of GlutaMAX, antibiotic–antimycotic, and Na-pyruvate in a 37°C water bath, rinse all tubes with 70% ethanol, and wipe dry.

2. Open a 500-ml bottle of DMEM without phenol red, Na-pyruvate, or glutamine and with low glucose (GIBCO/Invitrogen) in a clean, running laminar flow hood (see Note 5).

3. Add all tube contents into the DMEM. Addition of these ingredients does not significantly change the pH or the osmolality of the medium. GlutaMAX is a good substitution of glutamine with longer life in medium.

4. Pour approximately 30 ml medium into a 50-ml blue cap tube, and add 1,875 mg BSA (see Note 7) powder, cap the tube, and

vortex until dissolved. The mixture will generate foam, but it will not decrease the protein amount significantly (see Notes 6 and 7). In a running hood, filter the BSA solution directly into the DMEM bottle using sequentially a 0.45- and 0.22-μm filter connected to a 60-ml polypropylene (PP) syringe. Mix well by gentle inversion.

5. Write all added materials on the bottle, date, and store the medium at 4°C. Warm it to 37°C in a water bath before use. This is the base medium (A) to make the complete culture medium. We use this medium for 6 months.

3.2. Preparation of the Complete Medium (B) for Ovary Culture

3.2.1. Preparation of 10 ml Insulin, Transferrin, Selenium, and Linoleic Acid (ITS±)-100×

Mix 100 μg insulin, 6.25 mg transferrin, 6.25 μg selenous acid, 5.35 mg linoleic acid and 1.25 gm BSA (see Note 8), and volume to 10 ml with autoclaved MilliQ water. This solution will have a shelf-life of 1 year and should be stored at 4°C. One milliliter of this solution is added to 100 ml culture medium, and the final concentration of ingredients per ml is: 0.1 μg insulin, 6.25 μg transferrin, 6.25 ng selenous acid, 5.35 μg linoleic acid, and 5 mg BSA. The potency of insulin drops significantly in a more than year old supplement; hence, making a 10 ml supplement at a time is ideal. Fresh batch can be formulated easily and quickly from stock ingredients.

3.2.2. Preparation of DMEM with ITS± (Medium B)

A 12-well cluster plate needs 0.5 ml medium B per well and a 24-well cluster plate needs 0.3 ml medium B per well. Therefore, calculate the total volume of B needed plus 1 ml extra for all groups to be plated.

If several ovary cultures will be done within a week or two, then it is better to make 50–100 ml of B. Add 1 ml of ITS+ to medium A to make 100 ml medium B.

3.3. Preparation of GPR30 siRNA

We have partially cloned the hamster GPR30 cDNA (Accession No. DQ237895). Using this sequence information and Dharmacon's online siRNA design center; we have designed a series of putative hamster siRNA sequences. We have also used http://WWW.design.RNAi.jp (15) online software to design a series of GPR30 siRNA, and then selected the siRNA, which satisfied the following four conditions (15): (1) A/U at the 5′-end of the antisense strand, (2) G/C at the 5′-end of the sense strand, (3) AU-richness in the 5′-terminal third of the antisense strand, and (4) the absence of any GC stretch over 9 bp in length. The siRNA was synthesized by Dharmacon (Thermo-Fisher).

1. Dissolve the lyophilized siRNA (see Note 9) in autoclaved MilliQ water to a stock concentration of 20 μM. The final concentration will be 25, 50 or 100 nM depending on the optimal amount; hence, the stock should be stored in smaller aliquots at −80°C so that the leftover, if any, will be small and can be discarded.

3.4. Ovary Collection and Culture

3.4.1. Preparation of Pregnant Hamsters

Golden hamsters with 90–100 gm body weight should be kept at 14:10 light:dark cycle in comfortable quarter with food and water ad libitum. It is important to keep animals in a noise-free area. Animals should be checked for at least THREE consecutive estrous cycles (see Note 10).

1. Place a female in the cage of a male (100–130 gm body weight) at 4–5:00 p.m. on the 4th day of the estrous cycle, i.e., proestrus.

2. Check the vaginal lavage next morning for the presence of sperm, which denotes Day 1 of pregnancy. Hamster gestation lasts for 16 days with delivery in the noon/afternoon of 16th day.

3. Separate mated females from males. Pregnant females can be group-housed until about 12th day of pregnancy provided there is no fighting. In that case, females must be housed separately. Add a bedding material in the cage on the 12th day of pregnancy. A female can deliver 12–14 pups, but the number of female pups may vary considerably. The usual rule of thumb is to assume at least four female pups in the litter and calculate the number of pregnant hamsters needed for the experiment. If the number of female pups exceeds the expectation, additional groups can be added immediately to the experiment.

3.4.2. Collection of Fetal Ovaries

1. Turn on the laminar flow hood in the culture room, clean the bench with 70% ethanol, and wipe dry. Also, clean the tissue collection microscope table in the laboratory with 70% ethanol and wipe dry.

2. Presoak gauze in sterile PBS at room temperature by squirting from the bottle.

3. Take out all media to be used, and warm up the base medium for 15–20 min in a 37°C water bath. Pour 1–2 ml medium A in four sterile IVF culture dishes. Cover all dishes and keep them on the laminar flow bench with the hood running. IVF dishes have a secondary central chamber (Fig. 1a) to keep the ovaries in the center and to prohibit their movement to the periphery of the plate.

4. Take one of the dishes to the laboratory and keep near the dissecting scope.

5. In a separate place away from the tissue collection, preferably near a sink, place one 15-day pregnant female in the CO_2 chamber and allow the gas to enter without the hissing sound. Once the hamster is unconscious but still breathing, quickly inject 1 ml of 15% Nembutal (sodium pentobarbital) directly into the heart using a 25-gauge needle and a 3 ml disposable hypodermic syringe. This will euthanize the animal immediately without any stress (see Note 11).

6. Rinse the ventral surface of the animal with 70% ethanol by squirting from the bottle and wipe with a clean paper towel.

7. Place the animal on a clean paper towel under the dissecting scope.

8. After waiting 5 more minutes to ensure fetal euthanasia, cut open the abdomen with a pair of sharp scissors; take out one uterine horn with fetuses, and place it on a sterile PBS-soaked gauze and cover it with the same gauze. This will prevent tissue drying.

9. Cut open each fetal sac with a pair of iris scissors, remove the fetus by severing the umbilical cord and place it on dry paper towel under the scope. Fetal skin will stick to the paper and provide a natural restraint. Cut the abdominal skin and muscle with the iris scissors in a 'V' shape all the way to the side of the rib cage.

10. Locate the left sided ovary. It will be next to the kidney covered by the oviduct (Fig. 1b). Gently hold the uterine horn with the jewelers' forceps and cut. Gently pull the tissue away from the kidney and simultaneously cut the ligaments to free the ovary–oviduct structure. The tissue is extremely soft and can be torn easily. Therefore, gentle dissection is must.

11. Place the entire cut structure (Fig. 1b) in the medium in the dish. Repeat for the right side, and for subsequent fetuses. If testicles are needed for examination, collect those from male fetuses; otherwise, discard male fetuses.

12. When all fetuses are removed from the mother, do thoracotomy to ensure euthanasia. Keep record of male and female fetuses per pregnant female for USDA inspection. Wrap dead animals in paper towels, double bag them, and then dispose of according to institution's guidelines.

3.4.3. Culture of Fetal Ovaries

1. Bring all ovary–oviduct complexes in the hood and carefully dissect out ovaries from the oviduct (Fig. 2a). During dissection sort all ovaries to one side and all oviducts to other (Fig. 2b).

2. Using a sterile glass Pasteur pipette primed with the base medium remove all ovaries to a new dish with fresh medium at room temperature. Allow ovaries to disperse, and then gather them, and transfer to a second fresh medium. Repeat this process two more times to wash ovaries thoroughly with fresh medium.

3. Transfer ovaries to a new dish containing 2 ml medium B and repeat the process one more time.

4. Add 0.5 ml complete culture medium containing any test factor, such as estradiol-BSA conjugated with or without GPR30 siRNA in a 12-well tissue culture cluster.

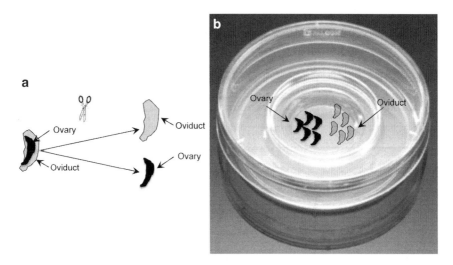

Fig. 2. Diagrams showing (**a**) separation of an ovary from the ovary–oviduct complex, and (**b**) arrangement of ovaries and oviducts in the IVF dish after separation.

3.4.4. Use of GPR30 siRNA in Culture

In the beginning, a trial run is needed to optimize the effective amount of siRNA and a volume of Metafectene (Biontex, GmbH) that is nontoxic to cells. It is often useful to use a FITC-labeled nontargeting RNA (Dharmacon) to determine the transfection efficiency and optimal volume of Metafectene. Both Metafectene and siRNA/FITC-RNA should be at room temperature.

1. The volume of our culture medium is 0.5 ml (12-well format); therefore, the volume of stock siRNA added to each culture well is: 0.625 μl (800-fold dilution to obtain a final concentration of 25 nM), 1.25 μl (400-fold dilution to obtain a final concentration of 50 nM), or 2.5 μl stock (200-fold dilution to obtain a final concentration of 100 nM). The following solution is made in a 96 well tissue culture plate with a concave base.

2. Label wells: A1–A4, B1–B4, and C1–C4.

3. Pipette 50 μl of medium B without serum (see Note 12) (we do not use serum in culture) in wells marked as indicated and add: 0.625 μl siRNA stock to A1–A4m; 1.25 μl siRNA stock to B1–B4, and 2.5 μl siRNA stock to C1–C4m. Mix the solution by pipetting one time; keep reagents inside the hood all the time.

4. Pipette 50 μl of medium B without serum in wells marked D1–D4, E1–E4, and F1–F4. Add the following Metafectene amount: 1, 2, 4, 6 μl in D1–D4, 1, 2, 4, 6 μl in E1–E4, and 1, 2, 4, 6 μl in F1–F4. Mix the solution by pipetting once. The amount of nucleic acid is very small even for the lowest volume

of Metafectene; hence, there is no need to use Metafectene volume higher than 6 µl. However, if the nucleic acid amount reaches micromol level, then larger volume of Metafectene must be included in the titration.

5. Using an Eppendorf micropipettor, combine well contents as follows: A1 + D1, A2 + D2, A3 + D3, A4 + D4. Do the same for the B and E, and C and F series. Do not mix.

6. Wait for 20 min; cover the plate with aluminum foil to avoid the photo bleaching of FITC.

7. Add the entire content of each well to the corresponding culture well (see Note 13) gently swirl the culture cluster to distribute the nucleic acid in the medium.

8. Place an insert in the well (Fig. 3a). Only the bottom of the insert should touch the surface of the medium in the well (Fig. 3a). Otherwise, the excess medium will force enter the insert through the wet area and completely submerge the ovaries (see Note 14).

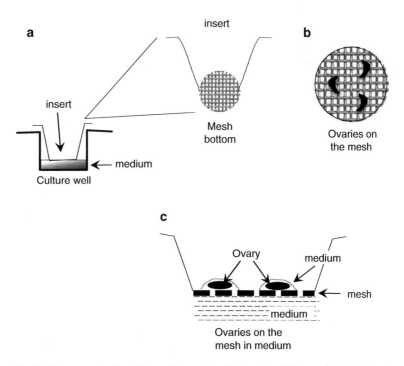

Fig. 3. Diagrams showing (a) a culture well with insert and the mesh of the insert, (b) placement of ovaries on the mesh, and (c) placement of medium on ovaries and insert on the medium.

9. Carefully place one ovary at a time (2 ovaries in 24-well or 3 ovaries in 12-well insert) with a miniscule amount of medium using a sterile glass Pasteur pipette (Fig. 3b). There should be just a thin film of medium covering the ovary. This ensures the proper exchange of O_2/CO_2. The ovaries will grow flatter with culture days, but will retain follicular morphology and functionalities (see Note 15). A proper placement of ovaries on the insert is shown in Fig. 3c.

10. Place the lid and transfer the entire plate to a humidified culture incubator containing 5% CO_2 in air.

11. Change the media (no siRNA) every 48 h over an 8 day period and terminate the culture on 9th day (see Note 13).

12. To ensure that ovarian phenotype is actually due to siRNA knockdown of target mRNA, and consequently of the protein product, molecular analysis of the protein level (see Note 16) must be evaluated. It is also necessary to analyze the protein product of a closely related gene for validation of the specificity of the siRNA effect. Therefore, before using for morphologic evaluation, use ovaries for protein extraction. Analyze the protein levels of the targeted and nontarget genes by Western blot in control and siRNA-treated ovaries. An example is presented in Fig. 7 (19).

13. Check ovaries and culture medium each time before medium replacement for tissue degradation and culture contamination. If these occur, culture should be discontinued and a fresh culture should be initiated with appropriate precautions.

3.4.5. Culture Termination and Processing of Ovaries

1. On 9th day, remove the insert from the culture well, and place it in a used culture cluster plate containing ice-cold PBS. Add cold PBS in the insert. Repeat the process twice to remove most of the protein in the culture medium.

2. Half-fill a glass Pasteur pipette with cold PBS; place the tip close to the ovary on the mesh, but do not touch the tissue. Force a jet of PBS to dislodge the tissue. Repeat the procedure several times until the ovary is dislodged from the mesh. Care must be taken not to aspirate the ovary in the pipette repeatedly; otherwise, the tissue will be fragmented.

3. Continue the process until all ovaries are dislodged (see Note 17).

4. Ovaries can be used for analysis of gene expression using real-time quantitative PCR, immunofluorescence or Western blot analysis of proteins, or morphological analysis of primordial follicle formation. For RNA or protein analysis, transfer ovaries with a small amount of PBS to 0.65-ml RNA–DNA-free polypropylene tubes. Aspirate the PBS using a small transfer pipette

under a stereomicroscope, and snap-freeze the tube contents in liquid N_2. Store ovaries at $-80°C$ until used. However, it is advisable not to store intact ovaries for a long time to avoid RNA degradation. For convenience, ovaries can be transferred to RNA-later (Ambion) solution in a 30 mm petri dish first, and then to the storage tube for snap-freezing. RNA-later prevents RNA degradation during tissue handling. Remove RNA-later from the tissue before freezing. For morphological analysis, transfer ovaries to a 12×75 mm glass tube containing at least 1 ml of Bouin's fixative.

5. Fix ovaries for at least 24 h, rinse 3× with room temperature PBS and dehydrate in 1 ml 70% ethyl alcohol in MilliQ water containing approximately 5 mg lithium carbonate (see Note 18) through 100% ethanol following the JB-4 manufacturer protocol. We process cultured ovaries in a used 12-well culture cluster placed on ice in a small Styrofoam box, with gentle rocking for 1 h in each grade of ethanol.

6. After the final ethanol step, place ovaries in JB-4 (EM Sciences, Hatfield, PA) infiltration solution (see Note 19), and follow the instruction provided by the manufacturer to make plastic blocks (Fig. 4a–c). Plastic blocks must be cut in a motorized microtome, such as Leica microtome, using a tungsten-carbide knife for even thickness.

7. Cut 5 μm thick tissue sections one section at a time and take them on small drops of deaerated (see Note 20) MilliQ water on a clean microscope slide coated with appropriate adhesive using plastic sectioning protocol. Usually, three rows of 5–6 section per row are optimal for a slide, and it avoids section overlaps and wrinkles (Fig. 5).

8. Stain sections using routine hematoxylin/eosin staining technique, and mount with DPX (dibutylphthalate, tricresyl phosphate, and xylene; EM Sciences).

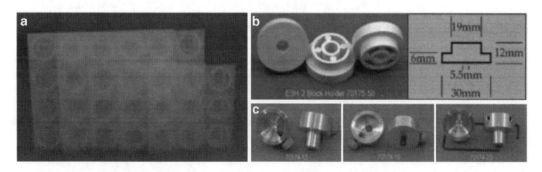

Fig. 4. Images of (**a**) molding cup trays, (EM Sciences, #70176-20), (**b**) EBH-2 block holder (EM Sciences, #70175-50), and (**c**) Aluminum chuck (EM Sciences, #70174-10/14/20).

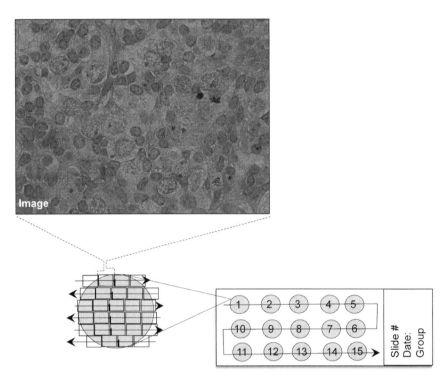

Fig. 5. Diagram showing the orientation of sections on a slide, and the direction of counting sections and images in a section, and a picture of H & E stained 8-day-old hamster ovary section.

3.5. Image Acquisition, Identification of Primordial Follicles in the Image, and Analysis

Good knowledge of microscopy and ovarian histology is essential for morphological analysis of primordial follicle formation.

1. Capture images from one side to another and from top to the bottom of the section as shown in Fig. 5 taking care not to overlap sections. It is useful to select an arbitrary visual landmark at the end of each image in the field of view to avoid the overlap. Continue from left to right until all sections were digitized (Fig. 5).

2. A primordial follicle is an oocyte completely surrounded by 1 or few flattened granulosa cells (3) (Fig. 6a, b). It is important to ensure that cells adjacent to the oocyte are truly apposed to the oocyte plasma membrane; otherwise, nongranulosa cells will be identified as granulosa cells and an unassembled oocyte will be scored wrongly as a primordial follicle. This can be done by slightly changing the focus. Cells truly apposed to the oocyte will change focus along with the oocyte, but cells that are just in the vicinity will change focus inversely.

3. Using Image pro Plus application mark oocytes and primordial follicles (Fig. 6a, b). Duplicate counting of the same oocytes or follicles can be avoided by one of the two ways:

   (a) If all consecutive sections will be used for counting, mark only oocytes presenting a nucleolus (Fig. 6b). The oocyte

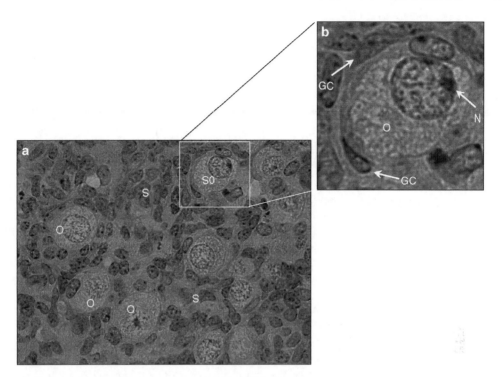

Fig. 6. Sections of an in vitro grown 15-day-old (E15) fetal hamster ovary treated with estradiol-17β for 8 days. A magnified view of a primordial follicle is also presented.

can be with or without granulosa cells. Because the nucleolus is smaller than 5 μm in diameter, the same nucleolus will not be present in the next section, and will eliminate duplicate counting of the same structure. This is the method of choice for counting when ovaries contain follicles at different stages of development.

(b) If all oocytes, regardless of the nucleolus, and primordial follicles in a section will be counted, then every fifth section should be used. Because the diameter of primordial oocytes is approximately 15 μm, same oocytes are unlikely to be present in every 5th section.

4. Count the total number of oocytes and primordial follicles present in all sections for each ovary and then calculate the percentage of primordial follicles with respect to the oocyte. In the hamster, primordial follicles appear first in an 8 day old ovary, and no other follicular structure is present. Therefore, their number cannot be calculated in terms of follicles. It is commonly done for ovaries containing follicles at different stages of development.

3.6. Staging Developing Follicles in Adult Ovaries

Follicular development has routinely been studied in juvenile and adult animals using ovaries after experimental manipulations, such as unilateral ovariectomy, exogenous gonadotropin or steroid hormone treatment, and so on (3). In the hamster, stages of follicular development are exquisitely defined (20). Continuous proliferation of granulosa cells moves follicles from 1 layer primary stage through stage 2 (2 layers GC), stage 3 (3 layers GC), stage 4 (4 layers GC), stage 5 (5–6 layers GC), and stage 6 (7–8 layers GC and a layer of thecal cells) (20) (Fig. 7a). Thereafter, small cavities (antrum) develop in the GC layers leading to the formation of stage 7 follicles (incipient antrum and well-developed thecal layer). This is an important transition in follicular life because it heralds follicular selection and critically depends on FSH action (3, 21, 22). With one or two more rounds of GC proliferation and expansion of the antral cavity, follicles move to stages 8 through 10, which are antral follicles with increasing size (20) (Fig. 7a). Stage 10 follicles become the Graafian or Preovulatory follicles under the influence of the Preovulatory LH surge (3). The staging of follicles in other species has been discussed (2, 23–25), but the fundamental principle remains similar.

(a) Collect ovaries from juvenile or adult hamsters, fix in Bouin's fixative, and process, section, and stain as described earlier.

(b) Capture images as described earlier and count follicles at different stages based on the criteria mentioned earlier (20). However, for these ovaries, only oocytes with nucleolus must be counted (25) to avoid duplicate counting.

(c) If the goal is to determine the absolute number of follicles or oocytes in an ovary, then sections from one end of the ovary to

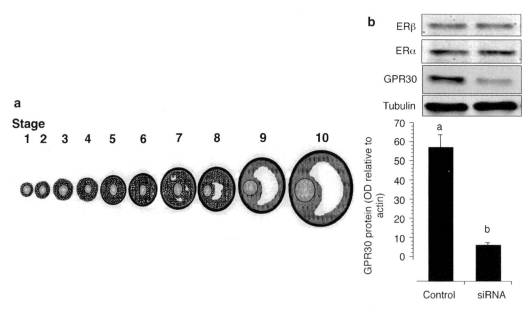

Fig. 7. (a) Diagrams of follicles at different stages of development, and (b) Western blot and digitization of ERa, ERb, and GPR30 proteins in 9 days cultured E15 hamster ovaries treated with GPR30 siRNA (19).

the other must be examined. However, for most studies this approach is unnecessary because the percentage of follicles at each stage relative to the total number of follicles is sufficient to detect any change in the dynamics of follicular development.

Count the number of follicles at each stage of development until the total number reaches 400–500 follicles. If desired, make a separate category for atretic follicles, which show pyknotic (dense purple to black) GC nuclei. We consider a follicle atretic if it has at least 2–3 pyknotic nuclei (3). Then, derive the percentage of follicles at each stage relative to the total number of follicles (26).

4. Notes

1. It is important to thoroughly clean the area of dissection and the laminar flow hood bench with 70% ethanol before and after each use. All media should be warmed to 37°C by placing them in a water bath, preferably at least 20 min before use. The bottles should be rinsed with 70% ethanol and wiped with sterile KimWipes before placing them in the hood.

2. Blood and tissue fluid are collected at the juncture of two scissor blades during use, and often bypass cleaning. Open the spring lock at the end, which allows the blades to be wide open for thorough cleaning. Relock the handles after cleaning.

3. The euthanasia procedure must be preapproved by the Institutional Animal Care and Use Committee (IACUC). We use a combination of CO_2 overdose and intracardiac injection of Nembutal to ensure death of the animals including the fetuses.

4. The animal use protocol must be preapproved by IACUC. The hamster is a USDA-regulated species; hence, additional animal handling precautions must be included in the protocol.

5. GlutaMAX, antibiotic–antimycotic, and Na-pyruvate solution come in 100 ml size. A new bottle should be thawed in a 37°C water bath, rinsed in 70% ethanol, wiped dry, and opened in a laminar flow hood bench with hood running. The liquid in the bottle should be divided into 5 ml aliquots in sterile 15-ml white cap PP tubes in the hood and stored at –20°C. Before use, each tube should be thawed in 37°C water bath, rinsed with 70% ethanol, wiped dry, and opened in the hood. This eliminates repeated freezing–thawing, which is detrimental to all solution.

6. Mouse and human ovaries are routinely cultured in McCoy 5 medium or α-MEM (27, 28).

7. The final BSA concentration of the culture medium is 0.5%. Added BSA makes the concentration 0.375%. The rest of the BSA comes from the ITS⁺ supplement.

8. A 30–40% BSA solution will be very viscous. That much BSA will displace a significant volume of water; and will take a long time to dissolve. Therefore, it is better to add the BSA to 70–80% of the final volume, and stir it gently over a magnetic stirrer at 4°C overnight. Next day, volume with autoclaved water, mix well with inversion and filter the solution first through a 0.45 μm MilliQ filter followed by a 0.22 μm filter using a 60 ml PP syringe in a laminar flow hood. It will take significant force to filter the BSA solution. Store frozen at −20°C in 5 ml aliquots.

9. siRNA is single stranded nucleic acid similar to mRNA; hence, all appropriate care for handling mRNA must be strictly followed to avoid degradation. Repeated freezing–thawing should also be avoided.

10. Hamster estrous cycle lasts for 4 days, and is highly sensitive to light–dark schedule. We use 14:10 light:dark cycle. Copious vaginal mucus is discharged on the day of estrus, which is called Day 1 of the cycle when ovulation occurs at 2:00 a.m. The stringy discharge has a characteristic pale-yellow color and strong sulfide-like odor. It decreases considerably or disappears in the afternoon; hence, it is advisable to check the estrous cycles by 10:00 a.m. each day. Because the discharge returns on each estrus, each animal is handled once in each 4-day estrous cycle, thus reducing the stress considerably. It is easy to learn and perfect checking the hamster estrous cycles from just one demonstration, which only needs visual observation of mucus discharge when the perineal area is pressed gently, and by touching the discharge for its stringiness. The entire process should take no more than 10–15 s. Each cage card will have many dates showing the consecutive nature of the cycle. We place 5–6 nonpregnant hamsters per cage according to UNMC Comparative Medicine policy.

11. Because animals rendered unconscious with CO_2 overdose can revive quickly when removed from the chamber, they should be injected immediately and quickly with Nembutal. However, care must be taken to inject the anesthetic only in the heart and not the adjacent area. A few practice runs with dead animals can help one to understand the correct location of the heart with relation to the chest wall. Intraperitoneal injection is not recommended because the injection will invariably be in the uterus with fetuses. This will not accomplish euthanasia,

but may lead to animal suffering, which must be avoided at every level.

12. There should be no serum in the medium when siRNA is added because serum interferes with transfection. The well should be washed twice with fresh serum-free medium and then siRNA added to an appropriate volume of serum-free medium in the culture well. OptiMEM (GIBCO/Invitrogen) can be used instead during siRNA transfection; however, it does contain higher levels of insulin and growth additives. Therefore, it is important to ensure that none of the additives introduces a confounding variable to the culture.

13. If siRNA is to be added later in the culture, then medium should be changed before siRNA administration. Otherwise, the addition should coincide with the plating of the culture, i.e., day 0 of culture.

14. It is extremely important that ovaries are not excessively submerged in medium on the insert or they will perish: approximately 1 mm thick layer of medium should be on top of the ovaries.

15. Ovaries will become flatter as the culture progresses and by 9th day, they will be very flat on the mesh, but will remain glossy and translucent without a black central core. A black core indicates dead cells, either due to poor culture condition, being excessively submerged due to media overflow in the insert, or due to experimental intervention.

16. One can determine the levels of both mRNA and protein of the targeted gene, but an examination of the protein level is critical because it reflects the biological effect. A decrease in the transcript levels by siRNA may not produce a desired end result, unless the decrease is accompanied by a markedly reduced translation product.

17. If the treatment condition causes ovaries to spread thin at the periphery, the peripheral layers of cells will not dislodge from the mesh because cells protrude through the pore. In most cases, this is not a problem because morphologic changes can be detected throughout the ovary. However, if this part of the ovary is important for the study, then membrane must be cut using a sharp needle and processed for downstream analysis.

18. Lithium carbonate removes picric acid from the tissue. It is important to remove as much of the picric acid as possible; otherwise, hematoxylin staining of the nuclei will be unsatisfactory.

19. JB-4 is a polymer embedding medium, which allows thin sections of ovaries with outstanding morphological details. Ovary sections can be stained with aqueous based hematoxylin and eosin. It is important to follow the JB-4 kit instruction

manual for proper infiltration and embedding to make blocks with excellent cutting property.

20. Vacuum deaeration of autoclaved MilliQ water is essential to avoid air bubble formation underneath the sections. The water must be at room temperature and deaerated each day before use.

Acknowledgment

The chapter is based on the work that was supported by grants R01-HD 38468 and Olson Foundation, Omaha to S. K. Roy. Cheng Wang was a post-doctoral research associate. Currently, he is a K99/R00 award (K99HD059985) recipient from the NICHD. Anindit Mukherjee and Prabuddha Chakraborty are graduate students conducting research on formation and development of primordial follicles.

References

1. Byskov, A.G. (1986) Differentiation of mammalian embryonic gonad. *Physiol Rev*, **66**, 71–117.

2. Hirshfield, A.N. (1991) Development of follicles in the mammalian ovary. *Internat.Rev. Cytol.*, **124**, 43–101.

3. Greenwald, G.S. and Roy, S.K. (1994) In Knobil, E. and Neill, J. D. (eds.), *The Physiology of Reproduction*. 2nd ed. Raven Press, New York, Vol. 1, pp. 629–724.

4. Eppig, J.J. (2001) Oocyte control of ovarian follicular development and function in mammals. *Reproduction*, **122**, 829–838.

5. Pepling, M.E. (2006) From primordial germ cell to primordial follicle: mammalian female germ cell development. *Genesis*, **44**, 622–632.

6. Skinner, M.K. (2005) Regulation of primordial follicle assembly and development. *Hum Reprod Update*, **11**, 461–471.

7. Roy, S.K. (1999) In Jay, K. P., Krishna, A. and Halder, C. (eds.), *Comparative Endocrinology and Reproduction*. Narosa Publishing House, New Delhi, pp. 315–330.

8. Edson, M.A., Nagaraja, A.K. and Matzuk, M.M. (2009) The Mammalian ovary from genesis to revelation. *Endocr Rev*, **30**, 624–712.

9. Aittomaki, K., Herva, R., Stenman, U.H., Juntunen, K., Ylostalo, P., Hovatta, O. and De la Chapelle, A. (1996) Clinical features of primary ovarian failure caused by a point mutation in the follicle-stimulating hormone receptor gene. *J Clin Endocrinol Metab*, **81**, 3722–3726.

10. Eckstein, F. (2005) Small non-coding RNAs as magic bullets. *Tredns in Biochemical Sciences*, **30**, 445–452.

11. Tang, G. (2005) siRNA and miRNA: an insight into RISCs. *Trends Biochem Sci*, **30**, 106–114.

12. Martinez, J., Patkaniowska, A., Urlaub, H., Lührmann, R. and Tuschl, T. (2002) Single-stranded antisense siRNAs guide target RNA cleavage in RNAi. *Cell*, **110**, 563–574.

13. Elbashir, S.M., Lendeckel, W. and Tuschl, T. (2001) RNA interference is mediated by 21- and 22-nucleotide RNAs. *Genes Dev*, **15**, 188–200.

14. Elbashir, S.M., Harborth, J., Lendeckel, W., Yalcin, A., Weber, K. and Tuschl, T. (2001) Duplexes of 21-nucleotide RNAs mediate RNA interference in cultured mammalian cells. *Nature*, **411**, 494–498.

15. Naito, Y., Yamada, T., Ui-Tei, K., Morishita, S. and Saigo, K. (2004) siDirect: highly effective, target-specific siRNA design software for mammalian RNA interference. *Nucleic Acids Res*, **32**, W124–129.

16. Ui-Tei, K., Naito, Y., Takahashi, F., Haraguchi, T., Ohki-Hamazaki, H., Juni, A., Ueda, R. and Saigo, K. (2004) Guidelines for the selection of highly effective siRNA sequences for mammalian and chick RNA interference. *Nucleic Acids Res*, **32**, 936–948.

17. Naito, Y., Yamada, T., Matsumiya, T., Ui-Tei, K., Saigo, K. and Morishita, S. (2005) dsCheck: highly sensitive off-target search software for double-stranded RNA-mediated RNA interference. *Nucleic Acids Res*, **33**, W589–591.

18. Wang, C. and Roy, S.K. (2006) Expression of growth differentiation factor 9 in the oocytes is essential for the development of primordial follicles in the hamster ovary. *Endocrinology*, **147**, 1725–1734.

19. Wang, C., Prossnitz, E.R. and Roy, S.K. (2008) G protein-coupled receptor 30 expression is required for estrogen stimulation of primordial follicle formation in the hamster ovary. *Endocrinology*, **149**, 4452–4461.

20. Roy, S.K. and Greenwald, G.S. (1985) An enzymatic method for dissociation of intact follicles from the hamster ovary: histological and quantitative aspects. *Biol Reprod*, **32**, 203–215.

21. Kumar, T.R., Wang, Y., Lu, N. and Matzuk, M.M. (1997) Follicle stimulating hormone is required for ovarian follicle maturation but not male fertility. *Nat Genet*, **15**, 201–204.

22. Dierich, A., Sairam, M.R., Monaco, L., Fimia, G.M., Gansmuller, A., LeMeur, M. and Sassone-Corsi, P. (1998) Impairing follicle-stimulating hormone (FSH) signaling in vivo: targeted disruption of the FSH receptor leads to aberrant gametogenesis and hormonal imbalance. *Proc Natl Acad Sci USA*, **95**, 13612–13617.

23. Gougeon, A. (1981) In Rolland, R., VanHall, E. V., Hillier, S. G., McNatty, K. P. and Schoemaker, J. (eds.), *Follicular maturation and ovulation*. Excerpta Media, Amsterdam.

24. Pedersen, T. and Peters, H. (1968) Proposal for a classification of oocytes and follicles in the mouse ovary. *J Reprod Fertil*, **17**, 555–557.

25. Knigge, K.M. and Leathem, J.H. (1956) Growth and atresia of follicles in the ovary of the hamster. *Anat Rec*, **124**, 679–707.

26. Chiras, D.D. and Greenwald, G.S. (1978) Acute effects of unilateral ovariectomy on follicular development in the cyclic hamster. *J Reprod Fertil*, **52**, 221–225.

27. Hovatta, O., Wright, C., Krausz, T., Hardy, K. and Winston, R.M.L. (1999) Human primordial, primary and secondary ovarian follicles in long-term culture: effect of partial isolation. *Hum.Reprod.*, **14**, 2519–2524.

28. Eppig, J.J. and O'Brien, M.J. (1996) Development in vitro of mouse oocytes from primordial follicles. *Biol Reprod*, **54**, 197–207.

Chapter 13

Microspread Ovarian Cell Preparations for the Analysis of Meiotic Prophase Progression in Oocytes with Improved Recovery by Cytospin Centrifugation

Teruko Taketo

Abstract

Female fertility is critically influenced by two events affecting oocytes during meiotic prophase progression: meiotic recombination between homologous chromosomes; and a major oocyte loss. It is technically challenging to examine these events, which take place in fetal and neonatal ovaries in mammals. Here, we describe a protocol for the preparation of dissociated ovarian cells and their spread onto histology slides. These preparations are suitable for cytogenetic, quantitative, and FISH analyses.

Key words: Mouse ovary, Oocyte, Meiosis, Cytospin centrifugation, Microspread cell preparation

1. Introduction

Microspread (or surface spread) preparation of testicular cells has been widely used to study the localization of various molecules in association with homologous chromosome pairing and synapses in mammalian spermatocytes (1). However, similar approaches are limited in regards oocytes, largely because the meiotic prophase is initiated and proceeds in fetal ovaries, and the number of oocytes is very small. Yet, the mechanisms involved in female meiosis are distinct from those in spermatocytes (2), and it is imperative to study the oocytes during the meiotic prophase progression for a full understanding of mammalian meiosis.

Wai-Yee Chan and Le Ann Blomberg (eds.), *Germline Development: Methods and Protocols*, Methods in Molecular Biology, vol. 825, DOI 10.1007/978-1-61779-436-0_13, © Springer Science+Business Media, LLC 2012

In addition to the basic mechanisms of meiotic prophase progression, three aspects of female meiosis particularly need to be addressed. First, there is a major loss of oocytes during the meiotic prophase progression, thus limiting the oocyte reserve for reproduction. The link between meiotic events and oocyte loss remains to be clarified. Second, it has recently been found that the onset of meiosis is regulated by retinoic acid and its metabolism; the germ cells can be induced to enter meiosis in fetal testes, but they eventually die by birth (3–5). It remains to be studied whether male germ cells go through meiotic pairing and recombination in fetal testes. Third, attempts have been made to induce oogenesis from stem cells in culture (6–9). It is important to demonstrate that they undergo proper meiotic events.

Various methods have been used to study oogenesis in fetal and neonatal ovaries. Histological sections are superb for capturing the distribution of germ cells and morphological events in vivo. However, it is difficult to estimate the number of germ cells, particularly at advanced stages when the size and stage of oocytes varies tremendously. It is also difficult to assess the meiotic events such as homologous pairing and recombination. Squashed preparations maintain the cells close to an in vivo state. However, the cells available for analyses are limited in number and do not represent the entire oocyte population. Dissociation of ovarian cells would improve the visibility of meiotic events as well as the representation. However, their recovery onto histology slides is troublesome. It is possible to let the cells settle on slides, but the majority of cells are flushed away during washing following fixation. The "dry-down" method compromises the resolution as well as the preservation of some proteins. We have introduced cytospin centrifugation to secure the maximum recovery of single cells onto histological slides (10). The total number of germ cells recovered on slides is smaller than those counted in histological sections. Nonetheless, this method is suitable for capturing the overall changes in oocyte populations and for comparing ovaries on different genetic backgrounds. In this chapter, we provide a full protocol for this procedure.

Our methods consist of two main procedures: (1) dissociation of ovarian cells and (2) their fixation onto histology slides by cytospin centrifugation. We have been using the chambers for cytospin centrifugation from Damon (Needham HTS, MA, Model IM-209) in our laboratory. Unfortunately, this model has been discontinued and is not easily available. Therefore, we describe an alternative method here using tissue culture chamber slides for centrifugation. The slides with microspread ovarian cells can be kept in a freezer and used for immunocytochemical labeling, RNA-FISH, or DNA-FISH. We routinely use hypotonic treatment of dissociated cells to expose nuclei, which enables excellent immunolabeling of nuclear proteins (10–12). Examples are shown in Fig. 1. We also describe an alternative method in which the cytoplasm is retained and therefore cytoplasmic proteins can be immunolabeled (11).

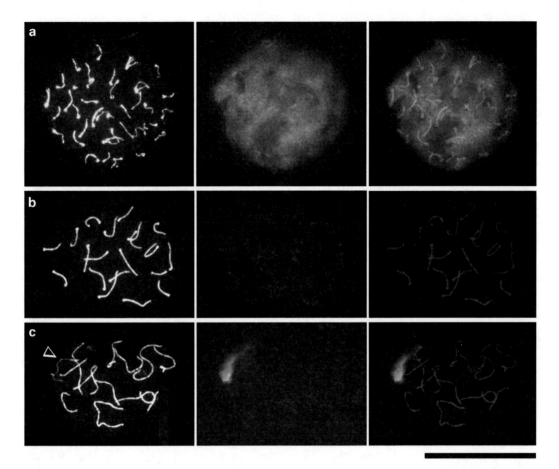

Fig. 1. Immunolabeling of microspread ovarian cells with the antibodies against synaptonemal complex proteins (*left*) and a phosphorylated histone variant γH2AX (*middle*), with merged images (*right*). (**a**) XX oocyte at the zygotene stage. Homologous chromosomes are partially synapsed while heavy γH2AX labeling of double-strand breaks (DSB) is seen in the entire nucleus. (**b**) XX oocytes at the pachytene stage. Discrete SC labeling is seen along 20 paired homologous chromosome cores. γH2AX labeling has disappeared due to the repair of DSB. (**c**) XO oocyte at the pachytene stage. The single X-chromosome is heavily labeled for γH2AX due to the lack of synapsis (*arrowhead*). Bar indicates 100 μm.

2. Materials

2.1. Dissociation of Ovarian Cells

1. Oven at 37°C.

2. Low-speed microcentrifuge.

3. MEM(H): Minimum Essential Medium with Hanks's salts (GIBCO) supplemented with 25 mM Hepes buffer (GIBCO). Store at 4°C. Warm up to room temperature before use.

4. 0.05% collagenase: Make fresh before use. Dissolve 1 mg of Type 1 collagenase (Worthington) in 2 ml of MEM(H). Keep on ice. Warm up to room temperature before use.

5. Rinaldini solution: 8 mg/ml NaCl, 0.2 mg/ml KCl, 1 mg/ml tri-Na citrate, 0.05 mg/ml $NaH_2PO_4 \cdot H_2O$, 1 mg/ml $NaHCO_3$, and 1 mg/ml glucose in double distilled H_2O (dd H_2O). Sterilize through a 0.2-μm filter. Store at 4°C.

6. 1.25% trypsin stock solution: 12.5 mg/ml trypsin (Worthington) in saline. Sterilize through a 0.2-μm filter. Divide into 100 μl aliquots in 1.5-ml microcetrifuge tubes. Store at –20°C.

7. 0.125% trypsin: Make fresh before use. Add 0.9 ml of Rinaldini solution to each tube containing 100 μl of 1.25% trypsin stock solution and mix. Keep on ice. Warm up to room temperature before use.

8. Trypsin inhibitor (TI): Make fresh before use. 20 mg/ml soybean trypsin inhibitor (Worthington) in MEM(H). Make just enough, 10 μl (containing 0.2 mg of TI) for each ovary. Keep on ice.

9. Heat-inactivated fetal bovine serum (FBS): Divide into 150 μl aliquots in 1.5-ml microfuge tubes. Store at –20°C.

10. 10% FBS: Make fresh before use. Add 1.35 ml MEM(H) to each tube containing 150 μl of FBS and mix. Keep on ice. Warm up to room temperature before use.

11. PBS: Sterilize through a 0.2-μm filter. Store at 4°C. Warm up to room temperature before use.

2.2. Microspread Nuclei Preparations

1. Low-speed centrifuge with swinging buckets for 94-well microplates.

2. 0.01 M borate buffer (pH 9.2): 3.81 mg/ml sodium borate·10H$_2$O in ddH$_2$O. Adjust pH to 9.2 with 0.5 N NaOH. Store at room temp.

3. 20% paraformaldehyde (PFA) stock solution: After opening, aliquot into microfuge tubes with minimum air and keep at 4°C. Use within a month (see Note 1).

4. 1% PFA: Make fresh before use (see Note 2). Dilute 20% PFA in ddH$_2$O to final 1%. Adjust pH to 8.2 with 0.01 M borate buffer. Keep at room temperature.

5. 0.5% NaCl hypotonic solution: Make fresh before use (see Note 2). Dissolve 5 mg/ml NaCl in ddH$_2$O. Adjust pH to 8.0 with 0.01 M borate buffer. Keep at room temperature.

6. 0.4% Photo-Flo: Make fresh before use (see Note 2). Dilute Photo-Flo 200 (Kodak) in ddH$_2$O to final 0.4%. Adjust pH to 8.0 with 0.01 M borate buffer. Keep at room temperature.

7. QuadriPERM cell culture vessel (SIGMA)

8. LabTek II CC^2 chamber slides (Thermo Scientific).

9. Plastic histology slide boxes.

10. Silica gel.

3. Methods

3.1. Dissociation
of Ovarian Cells

1. Isolate ovaries from adjacent mesonephroi and surrounding tissues in MEM(H) (see Note 3).

2. Transfer each ovary with a small volume of MEM(H) to a 1.5-ml microcentrifuge tube. Keep at room temperature.

3. After finishing the dissection of all ovaries, remove as much medium as possible with a Pasteur pipette and add 100 µl 0.05% collagenase to each tube.

4. Incubate the tubes in an oven at 37°C for 15 min.

5. Remove the collagenase solution and add 100 µl of MEM(H) to each tube.

6. Remove MEM(H) and add 100 µl of 0.125% trypsin to each tube.

7. Incubate the tubes in an oven at 37°C for 10–12 min (see Note 4).

8. Take the tubes out of the oven, and add 10 µl of 20 mg/ml trypsin inhibitor to each tube followed by gentle mixing as quickly as possible (see Note 5).

9. Remove the trypsin solution and add 100 µl 10% FBS to each tube. Gently mix. Avoid touching the ovaries during the pipetting (see Note 6).

10. Remove FBS and add 100 µl of MEM(H) to each tube.

11. Remove MEM(H) and add 100 µl of fresh MEM(H) to each tube.

12. Remove MEM(H) and add 100 µl of PBS to each tube.

13. Remove PBS and add fresh 200 µl of PBS to each tube. Pipette the ovary in and out of a siliconized yellow micropipette tip repeatedly (10–20 times) until no chunks of tissue are seen (see Note 7).

14. Centrifuge the tubes at $360 \times g$ for 3 min at room temperature.

15. Remove as much of the supernatant as possible with a micropipettor and resuspend the pellet in 20 µl of MEM(H). Keep at room temperature.

3.2. Microspread
Nuclei Preparations

1. Write down the sample information on LabTek chamber slides.

2. Place the chamber slides in QuadriPERM cell culture vessels.

3. Fill each chamber with 800 µl of 0.5% NaCl hypotonic solution. Select chambers in considering weight balance for centrifugation.

4. Add cell suspension (see Note 8) into the hypotonic solution, gently pipette up and down to spread the cells in the entire area, and let the chambers sit horizontally for 5 min (see Note 9).

5. Place the cell culture vessels in a swinging bucket rotor and centrifuge at $300 \times g$ for 5 min.

6. Slowly remove the hypotonic solution from a corner of each chamber with a Pasteur pipette or drain onto paper towel (see Note 10).

7. Immediately add 800 µl of 1% PFA to a corner of each chamber and let the chambers sit horizontally for 3 min.

8. Centrifuge at $100 \times g$ for 3 min.

9. Remove the fixative.

10. Repeat steps 7–9.

11. Add 800 µl of 0.4% Photo-Flo to each chamber.

12. Centrifuge at $100 \times g$ for 1 min.

13. Remove the solution.

14. Repeat steps 11–13 twice.

15. Take the chamber slides out and remove the chambers from the slides.

16. Let the slides dry under vacuum (10–15 min).

17. Place the slides in a plastic box containing silica gel (wrapped in tissue paper) and seal the box with a plastic tape. Store at –20°C (see Note 11).

3.3. Dissociated Whole Cell Preparations

After step 14 in Subheading 3.1

1. Remove all but a small volume of the supernatant, and gently add 100 µl of 1% PFA to each tube. Do not disturb the pellet (see Note 12). Let tubes sit for 3 min.

2. Centrifuge the tubes at $360 \times g$ for 3 min at room temperature.

3. Repeat steps 1–2.

4. Remove the fixative and add 100 µl of PBS.

5. Centrifuge the tubes at $360 \times g$ for 3 min at room temperature.

6. Remove as much of the supernatant as possible, and resuspend the pellet in 20 µl of MEM(H).

7. Prepare chamber slides as Subheading 3.2, steps 1 and 2.

8. Fill each chamber with 800 µl of PBS.

9. Add the cell suspension to PBS, gently pipette up and down to spread the cells in the entire area, and let the chambers sit horizontally for 5 min.

10. Centrifuge at $300 \times g$ for 5 min.

11. Remove the solution.

12. Immediately add 800 μl of 0.4% Photo-Flo to each chamber.
13. Centrifuge at $100 \times g$ for 1 min.
14. Remove the solution.
15. Repeat steps 12–14 twice.
16. Follow steps 15–17 in Subheading 3.2.

4. Notes

1. Homemade 1% PFA works if freshly made. However, the home-made solution loses its power due to oxidization after storage. For convenience, we purchase a 20% PFA solution in vials sealed with N_2 gas (Electron Microscopy Sciences). The solution is stable until opened. We have confirmed that the aliquots kept at 4°C work well if contact with air is minimized. We discard the aliquots if they are left in air for more than a couple of days or after 1 month of storage.

2. We routinely prepare 1% PFA, 0.5% NaCl, and 0.4% Photo Flo before use. We have never tested whether or not they can be stored.

3. The total number of ovaries to be dissected and processed at a time must be limited to the number of available chambers on slides for two reasons. First, the time of trypsin treatment is critical. If it takes too long, some nuclear proteins may be digested, producing inconsistent results. Second, it is undesirable to leave cell suspensions while the first set of cell preparations are processed through the entire procedure of cytospin centrifugation.

4. The time duration for incubation depends on the developmental stage of ovaries. We routinely incubate ovaries obtained at 14 days postcoitum (dpc) or earlier for 10 min and older ovaries for 12 min. However, the incubation duration may need adjustment (see Note 7).

5. When only one or two samples (e.g., ovaries are pooled) are processed, trypsin inhibitor is unnecessary. Replacement of trypsin inhibitor with 10% FBS is sufficient to stop trypsin digestion. When more samples are processed, trypsin continues to digest proteins until it is replaced with 10% FBS. Therefore, it is essential to stop the trypsin activity with its inhibitor as quickly as possible.

6. After the treatment with trypsin, ovaries become fluffy and are easily moved by suction force. It is important not to disturb the

tissue. Otherwise, the cells will dissociate prematurely. Repeated centrifugation is undesirable for the consistency of results.

7. When tissues are dissociated in PBS by pipetting, one can determine how trypsin treatment worked: if ovaries dissociated too quickly, they had been overdigested; on the contrary, if they do not dissociate after pipetting in and out 20 times, they have been underdigested. Time or concentration of trypsin treatment must be adjusted accordingly. It is also important to pipette in and out near the mouth of the plastic pipette tip. Use siliconized micropipette tips to minimize the loss of cells whenever cell suspension is pipetted.

8. When the total number of germ cells is to be counted, apply the entire cell suspension from each ovary into a chamber. If the number is too large (e.g., at early developmental stages), aliquots can be applied. However, the deviation of values appears to increase. This does not matter when only a portion of the germ cells is to be examined. The aliquots from an ovary can be applied to separate chambers and processed for different experiments.

9. The time duration for hypotonic treatment may need to be adjusted according to resultant chromosome spreading.

10. Rapid removal of solution may cause detachment of cells from the slides.

11. It is critical to keep the slides dry. After some slides have been taken out from the box, dry the remaining slides in the box containing silica gel under vacuum before sealing. It is better to process the slides as soon as possible. Some proteins (e.g., GCNA1) are stable, while others (e.g., CREST) are unstable during storage.

12. It is somehow more difficult to spin down single cells in fixative. It is better to keep the pellet undisturbed during fixation.

Acknowledgements

We would like to acknowledge Drs. H. Merchant-Larios (University of Mexico) and P. Moens (deceased) for inspiring us to develop the methods described here. We also thank our previous graduate students, especially Kelly McClellan and Michele Alton, who contributed to developing and applying the described methods. We finally thank Dr. P. Burgoyne and the members of his division for allowing me to test the alternative methods described here.

References

1. Moens PB (1994) Molecular perspectives of chromosome pairing at meiosis. BioEssays **16**:101–106.
2. Morelli MA, Cohen PE (2005) Not all germ cells are created equal: Aspects of sexual dimorphism in mammalian meiosis. Reproduction **130**:761–781.
3. Bowles J, Knight D, Smith C, Wilhelm D, Richman J, Mamiya S, Yashiro K, Chawengsaksophak K, Wilson MJ, Rossant J, Hamada H, Koopman P (2006) Retinoid signaling determines germ cell fate in mice. Science **312**:596–600.
4. MacLean G, Li H, Metzger D, Chambon P, Petkovich M (2007) Apoptotic extinction of germ cells in testes of Cry26b1 knockout mice. Endocrinology **148**:4560–4567.
5. Suzuki A, Saga Y (2008) Nanos2 suppresses meiosis and promotes male germ cell differentiation. Genes Dev **22**:430–435.
6. Hubner K, Fuhrmann G, Christenson LK, Kehler J, Reinbold R, De La Fuente R, Wood J, Strauss JF 3 rd, Boiani M, Schöler HR (2003) Derivation of oocytes from mouse embryonic stem cells. Science **300**:1251–1255.
7. Dyce PW. Wen L, Li J (2006) *In vitro* germline potential of stem cells derived from fetal porcine skin. Nature Cell Biol **8**:384–390.
8. Danner S, Kajahn J, Klink E, Kruse C (2007) Derivation of oocyte-like cells from a clonal pancreatic stem cell line. Mol Hum Reprod **13**:11–20.
9. Qing T, Shi Y, Qin H, Ye X, Wei W, Liu H, Ding M, Deng H (2007) Induction of oocyte-like cells from mouse embryonic stem cells by co-culture with ovarian granulosa cells. Differentiation **75**:902–911.
10. McClellan KA, Gosden R, Taketo T (2003) Continuous loss of oocytes throughout meiotic prophase in the normal mouse ovary. Dev Biol **258**:334–348.
11. Alton M, Taketo T (2007) Switch from BAX-dependent to BAX-independent germ cell loss during the development of fetal mouse ovaries. J Cell Sci **120**:417–424.
12. Alton M, Lau MP, Villemure M, Taketo T (2008) The behaviour of the X- and Y-chromosomes in the oocyte during meiotic prophase in the B6.Y^{TIR} sex-reversed mouse ovary. Reproduction **135**:241–252.

Chapter 14

In Vitro Maturation (IVM) of Porcine Oocytes

Ye Yuan and Rebecca L. Krisher

Abstract

Oocyte maturation is a critical component of in vitro embryo production. If not carried out in a precise manner under optimal conditions, subsequent fertilization and embryo development will be compromised. Here, we describe collection and in vitro maturation procedures in swine that maintain oocyte competence, resulting in successful embryo development following fertilization. These procedures can be used both for basic research purposes and large-scale production of mature oocytes for use in subsequent assisted reproductive technologies.

Key words: Oocyte, Porcine, In vitro maturation, Meiosis, Culture medium

1. Introduction

Offspring have been produced using in vitro maturation, fertilization, and culture of pig oocytes (1–6). However, not all oocytes placed into an in vitro maturation system are capable of developing to the blastocyst stage and subsequently resulting in pregnancy (7). The in vitro environment must support both nuclear and cytoplasmic maturation. While nuclear maturation, which entails the reinitiation and successful completion of meiosis I, can be easily assessed, cytoplasmic maturation is a much more complex, multifaceted process that is difficult to measure, but equally critical to successful oocyte competence and development. Even oocytes that complete nuclear maturation may be incompetent for further development following fertilization (8–10). Unfortunately, there is currently no method to measure completion of cytoplasmic maturation other than successful fertilization and embryonic development. Thus, careful attention to the culture environment during oocyte maturation is critical to optimize oocyte developmental potential.

Wai-Yee Chan and Le Ann Blomberg (eds.), *Germline Development: Methods and Protocols*, Methods in Molecular Biology, vol. 825, DOI 10.1007/978-1-61779-436-0_14, © Springer Science+Business Media, LLC 2012

The procedures described herein represent the best available technology to support maturation given our current understanding of oocyte physiology. It is likely that as we gain additional knowledge of what constitutes oocyte competence and cytoplasmic maturation, these procedures will evolve and continue to improve, resulting in better oocyte quality and higher percentages of embryo development.

Importantly, in vitro oocyte maturation is the foundation for multiple assisted reproductive technologies in swine, including nuclear transfer and transgenesis. Application of these technologies could positively impact agricultural food production, preservation of genetic diversity, and transgenic animal production, which has the potential for improving food animal performance characteristics and disease resistance, as well as the production of human pharmaceutical proteins, models for biomedical research, and a source of tissues for xenotransplantation. However, without efficient maturation systems, these technologies cannot realize their full potential.

2. Materials

2.1. Culture Media

All concentrations given (mM) are those present in the final working solutions. Water used for all media is 18.2 mΩ and TOC less than 50 ppb. Reagents are weighed on an analytical balance. All disposable, sterile plasticware is from BD Falcon, unless otherwise specified. Final filtration of all stock solutions, as well as preparation of working solutions, is performed using sterile technique under a biological safety cabinet or laminar flow hood to maintain sterility. All glassware used in preparation of media solutions are dedicated for embryo use, and are heat sterilized in an oven at 250°C for a minimum of 8 h. After use, glassware is immediately rinsed, then washed seven times with reverse osmosis (RO) water and seven times with milliQ water, then dried completely and covered with aluminum foil before subsequent resterilization.

2.1.1. Synthetic Oviductal Fluid (SOF)-4-(2-Hydroxyethyl)-1-Piperazineethanesulfonic Acid (HEPES) Bench Medium (11, 12)

Combine 100 mL SOF base stock, 75 mL glucose stock, 84 mL HEPES stock, 10 mL nonessential amino acids without glutamine (NEAA, 100×, MEM; MP Biomedicals), 20 mL essential amino acids without glutamine (EAA, 50×, MEM; MP Biomedicals), 0.336 g NaHCO$_3$ (4.0 mM), 0.036 g Na-pyruvate (0.33 mM), 0.2703 g l-Lactate (3.0 mM; MP Biomedicals, LLC), 0.2172 g alanyl-glutamine (Ala-Gln, 1.0 mM), 1 mL gentamicin (50 µg/mL), 1 g fraction V BSA (0.1% w/v). Fraction V BSA may be substituted with 0.1 g polyvinyl alcohol (PVA) to make SOF-HEPES-PVA, a completely defined medium. QS to 1,000 mL with ultrapure water, adjust pH to 7.35 and filter (0.22 μm). Store at 4°C and use within 1 month (see Note 1).

Stock solutions required for this medium are as follows:

(a) *SOF Base Stock*: 58.270 g NaCl (99.70 mM), 5.340 g KCl (7.16 mM), 1.62 g KH_2PO_4 (1.19 mM), 0.996 g $MgCl_2$-$6H_2O$ (0.49 mM), 1.891 g $CaCl_2$ (1.71 mM). Prepare in 1000 mL ultrapure water and filter in 0.22 μm 1 L Stericup GV (Millipore) (see Note 2). Store at 4°C and use within 1 month.

(b) *Glucose Stock*: 3.6 g Glucose (1.5 mM). Prepare in 1,000 mL ultrapure water and filter in 1 L Stericup GV. Store at 4°C and use within 1 month.

(c) *HEPES Stock*: 5.958 g HEPES (21 mM), 0.19 g Phenol red (0.19 mM). Prepare in 1,000 mL ultrapure water and filter (0.22 μm). Store at 4°C and use within 3 months.

2.1.2. Tissue Culture Medium (TCM)-199 Maturation Medium

TCM-199 Medium (Gibco/Invitrogen Corp.) is commercially available in both liquid and powder forms (see Notes 3 and 4).

Defined: 9.5 mL TCM-199, 100 μL glucose stock, 100 μL pyruvate stock, 100 μL cysteine stock, 100 μL PSA (penicillin, streptomycin, and amphotericin, 1%; MP BioMedicals; see Note 5), 50 μL luteinizing hormone (LH, 0.01 U/mL; Sioux Biochemical), 50 μL follicle-stimulating hormone (FSH, 0.01 U/mL; Sioux Biochemical), 10 μL epidermal growth factor (EGF, 10 ng/mL) (13–15), 0.01 g PVA (see Notes 6 and 7). Add reagents to TCM-199. Filter (0.22 μm, Millipore Millex GV).

Undefined: 8.5 mL TCM-199, 100 μL glucose stock, 100 μL pyruvate stock, 100 μL cysteine stock, 100 μL PSA, 50 μL LH (0.01 U/mL), 50 μL FSH (0.01 U/mL), 10 μL EGF (10 ng/mL), 1.0 mL porcine follicular fluid (pFF) (16) (see Note 8). Add reagents to TCM-199. Filter (0.22 μm, Millipore Millex GV).

Stock solutions required for this medium are:

(a) *Pyruvate Stock*: 0.05 g Na-pyruvate (0.91 mM final concentration) in 5 mL TCM-199. No need to filter. Store at 4°C and use within 1 week.

(b) *Cysteine Stock*: 0.05 g cysteine (0.5 mM final concentration) in 5 mL TCM-199. No need to filter. Store at 4°C and use within 1 week.

(c) *Glucose Stock*: 0.5496 g glucose (3.05 mM final concentration) in 10 mL TCM-199. No need to filter. Store at 4°C and use within 1 month (see Note 9).

2.1.3. Purdue Porcine Medium (PPM) for Maturation (17)

1972.5 μL ultrapure water, 500 μL PPM Base, 500 μL bicarbonate, 500 μL glucose, 500 μL l-lactate, 500 μL taurine, 50 μL ala-gln, 50 μL cysteamine, 50 μL cysteine, 50 μL citric acid, 50 μL

NEAA, 50 μL EAA, 50 μL MEM vitamins (MP Biomedicals), 25 μL LH (0.01 U/mL), 25 μL FSH (0.01 U/mL), 50 μL EGF (100 ng/mL), 2.5 μL ITS (0.5 μg/ml insulin, 0.275 μg/ml transferrin, 0.25 ng/ml sodium selenite), 50 μL recombumin (G-MM; Vitrolife), 10 mg fetuin. Prepare this solution (total volume 5 mL) and filter (0.22 μm).

Stock solutions required for this medium are:

(a) *PPM Base*: 5.844 g NaCl (100 mM), 0.3728 g KCl (5.0 mM), 0.0681 g KH_2PO_4 (0.5 mM), 0.1887 g $CaCl_2$ (1.7 mM), 0.2933 g $MaSO_4 \cdot 7H_2O$ (1.19 mM), 0.0446 g alanine (0.5 mM), 0.1501 g Glycine (2.0 mM), 0.005 g phenol red. Prepare in 100 mL ultrapure water and filter (0.22 μm). Store at 4°C and use within 1 month.

(b) *Supplement Stocks*: each stock should be made individually with 10 mL ultrapure water. No need to filter. Store at 4°C.

- *Bicarbonate*: 0.21 g $NaHCO_3$ (25 mM), 1 week expiration

- *Glucose*: 0.0360 g glucose (2 mM), 1 month expiration (17–19)

- L-*lactate*: 0.0541 g L-Lactate (6 mM), 1 month expiration

- *Pyruvate*: 0.0022 g Na-pyruvate (0.2 mM), 1 week expiration

- *Aln-gln*: 0.2172 g aln-gln (1 mM), 1 week expiration

- *Taurine*: 0.626 g taurine (5 mM), 1 month expiration (20)

- *Cysteamine*: 0.0386 g cysteamine (0.5 mM), 1 week expiration, (21)

- *Cysteine*: 0.1 g cysteine (0.57 mM), 1 week expiration, (22, 23)

- *Citric Acid*: 0.0961 g citric acid (0.5 mM), 1 week expiration

2.1.4. NCSU23 Maturation Medium (24)

Combine 10% pFF, 0.01 U/mL LH, 0.01 U/mL FSH, 0.57 mM cysteine, 10 ng/mL EGF in NCSU23 (25) and filter (0.22 μm) (see Note 10).

Stock solutions required for this medium are:

(a) NCSU23: 0.635 g NaCl (108.73 mM), 0.0356 g KCl (4.78 mM), 0.0162 g KH_2PO_4 (1.19 mM), 0.0189 g $CaCl_2$ (1.7 mM), 0.0294 g $MaSO_4 \cdot 7H_2O$ (1.19 mM), 0.2106 g $NaHCO_3$ (25.07 mM), 0.0217 g Ala-Gln (1 mM), 0.1 g glucose (5.55 mM), 0.0876 g taurine (7 mM), 0.001 g phenol red. Prepare in 100 mL ultrapure water and filter (0.22 μm). Store at 4°C and use within 1 month.

**2.2. Solutions
and Reagents**

1. Saline: Prepare 0.9% saline for ovary transport and washing by adding 9 g of NaCl to 1 L RO water. Filtration is not necessary.

2. Acid Alcohol Fixative: Mix 30 mL 95% ethanol, 15 mL acetic acid, and 5 mL chloroform (6:3:1) (see Note 11).

3. Acetoorcein Stain: Dissolve 100 mg orcein in a solvent composed of 4 mL acetic acid in 6 ml water. Filter (0.22 μm). Acetoorcein stain can be stored indefinitely at room temperature (see Note 12).

4. Mountant (posting material): Melt equal quantities of paraffin wax and Vaseline together, place into a 3-cc syringe, let solidify, and store indefinitely at room temperature. The ratio of wax and Vaseline can be adjusted to meet individual preferences (more or less firm).

3. Methods

3.1. Day 1

3.1.1. Medium Preparation

1. Make IVM medium the morning of porcine oocyte collection (see Note 13). Media and culture dishes are prepared using sterile technique under a biological safety cabinet or laminar flow hood to maintain sterility.

2. Maturation can be conducted in bulk or in drops.

   (a) Bulk: Use Nunc four well plates (Nunclon; Nalge Nunc Intl.). Place 500 μL of IVM medium into each Nunc well and cover with 400 μL mineral oil. One well is used for washing. No more than 50 oocytes are matured per well.
   or
   (b) Drops: Use 60×15 mm dishes (see Note 14). For maturation, the total drop volume is 50 μL, and 8–10 oocytes are matured per drop. Place 25 μL drops of IVM medium directly on the dish, cover with 10 mL mineral oil (see Note 15), and then add another 25 μL IVM medium to each existing drop to bring the total drop volume to 50 μL. In addition, prepare a wash dish, with three 250 μL drops (125 μL, cover with oil, add an additional 125 μL). This drop size is sufficient to completely wash the oocytes from the SOF-HEPES searching medium into the appropriate maturation medium with no carryover.

3. Place the prepared IVM medium plates in the incubator for at least 4–6 h before adding oocytes, for gas and temperature equilibration.

4. Warm 100 mL SOF-HEPES and 1 L saline in a 37°C warm oven.

3.1.2. Ovary Transportation

1. Porcine ovaries are collected at local abattoir and transported to the laboratory as soon as possible in warm (37°C) saline (see Notes 16 and 17).

3.1.3. Aspiration

1. Take the temperature of the ovaries when they arrive in the laboratory by placing a thermometer into the thermos for 2 min. Ovaries should be processed no later than 6 h postmortem for best results, and the temperature should not fall below 28°C.

2. Wash the ovaries well with warmed saline, until the blood is washed away. Transfer the ovaries from the thermos into a plastic colander, and then pour 500–1,000 mL warmed saline (37°C) over the ovaries while stirring them with your gloved hand. Then place them into a large, sterile (autoclaved) glass or plastic beaker and fill the beaker with fresh, warm (37°C) saline to cover the ovaries. These beakers are not dedicated to embryo use. Immediately after use, beakers are rinsed, then washed with laboratory soap, rinsed ten times with RO water, dried and covered with aluminum foil prior to resterilization. The colander is rinsed thoroughly with RO water after each use, and air dried.

3. Place the beaker on top of a slide warmer or in a water bath (both set to 37°C) to keep the ovaries warm. If several beakers are filled, hold extra beakers in the warm oven until use. Aspiration of ovaries should be done in a dedicated area located away from the location where oocytes and embryos are handled.

4. Aspirate 3–8 mm diameter (26) antral follicles from the ovary with an 18-gauge needle. The needle can be affixed to a 10-mL disposable syringe or a vacuum pump (Fig. 1) to provide suction.

5. Pierce the surface of the ovary just outside the visible boundary of the follicle, such that the needle enters the follicle under the surface of the ovary. The bevel of the needle should be pointing away from the aspirator. These precautions prevent leakage of follicular fluid and loss of the oocyte. While aspirating the fluid, use the needle to scrape the follicle walls to ensure the oocyte is recovered.

6. Collect the fluid in a 50-mL conical tube, which is kept in a metal block on top of the slide warmer or in a water bath. Each ovary should yield approximately three selected oocytes; peripubertal gilt ovaries may yield twice this on average. A pellet consisting of cumulus–oocyte complexes and granulosa cells should be visible at the bottom of the tube following aspiration. Up to 35 mL of follicular fluid is typically collected per tube (see Note 18).

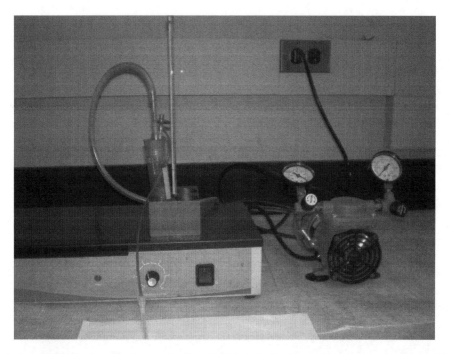

Fig. 1. Aspiration workstation for extracting oocytes from abattoir derived swine ovaries using a vacuum pump. The collection tube is attached to the pump via Tygon tubing (1/4-in. I.D., 3/8-in. O.D.) containing a 65 mm 0.2 μm PTFE membrane filter in the vacuum line (not visible in this photo) to protect the pump from accidental fluid aspiration. The tubing is then connected to an aspiration stopper using the barrel of a 1 cc plastic syringe cut in half. The aspiration stopper is composed of a silicon test tube stopper, with 2 18 gauge, 1.5 in. needles and 0.062-in. I.D., 0.125-in. O.D. tubing (Dow Corning medical grade silastic) passing through the stopper. One of the needles is attached to the Tygon tubing providing suction, as described. The second needle is attached to a 0.2 μm syringe filter to filter the air that flows into the tube. The tubing, approximately 18 in. in length, extends into the interior of the 50 mL collection tube about 1 inch. At the opposite end, there is a plastic connector (male luer lock to barb, Value Plastics, Inc) that fits into the tubing and attaches to an 18 gauge, 1.5 in. needle on the side opposite the tube. This is the needle that is used to aspirate ovarian follicles. The entire tube apparatus is then placed into a metal warming block on a slide warmer, which maintains the follicular aspirate at 37°C. Vacuum pressure is adjusted (typically less than 5 cm Hg) so that the aspirate moves slowly but steadily through the external tubing and into the collection tube. Care must be taken not to let the fluid level reach the vacuum intake or tubing on the inside of the stopper.

3.1.4. Oocyte Searching and Initiation of Maturation (Fig. 2) (See Note 19)

1. Using a sterile, disposable, individually wrapped transfer pipette (Fisher Scientific), aspirate approximately 1 mL of the pellet and place this into a square 150×15 mm grid dish (BD Falcon, 35–1112 Integrid) containing SOF-HEPES. Swirl to mix.

2. Search for oocytes under a stereomicroscope, visualizing one square of the grid dish per field of view (see Note 20). When oocytes are found, move them into a 60×15 mm petri dish of SOF-HEPES using a 5 μL handheld positive displacement pipette with glass bore tip (Drummond) (see Note 21). This collection dish should remain on the slide warmer at 37°C (see Note 22). While you are searching for COCs, another, smaller pellet may accumulate in the 50 mL tube. Aspirate this pellet into a second grid dish and repeat the search.

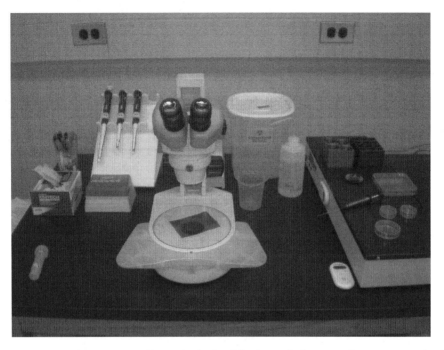

Fig. 2. Workstation set up for oocyte searching and selection following aspiration, as well as postmaturation. The slide warmer should be within easy reach of the stereomicroscope on the right side (left side for left-handed laboratory workers), for ease of transfer from search dish to selection dish using the micropipette (shown here on slide warmer). Temperature of the slide warmer is calibrated initially with a probe inside a medium dish, then monitored with a surface thermometer, pictured. Everything necessary for the work should be within easy reach to aid in working quickly. Trash is stored in Ziploc bags in a sealed plastic container (visible at right rear), and disposed of immediately after completing the work to prevent laboratory contamination. Used pipetmen and Drummond tips are collected in a plastic cup (mid table, right) and immediately disposed of after each work session.

3. When all the oocytes have been collected in the same 60×15 mm dish, increase the magnification to approximately 40× and examine each oocyte in the dish.

4. Select oocytes with even cytoplasm that are completely covered with several layers of compact cumulus cells. Do not select denuded oocytes, those with expanded or partially expanded cumulus cells or small oocytes (Fig. 3). Move the selected oocytes from the 60×15 mm dish through two 35×10 mm petri dishes (Falcon 1008) of warm SOF-HEPES, also kept on the slide warmer.

5. Remove the maturation dishes from the incubator. Place oocytes from the final HEPES wash into the maturation medium wash well or wash drop. Swirl to wash, then place into final maturation wells or drops. Oocytes are matured hours in 6% CO_2 balance air in a 100% humidified environment at 38.5°C in the incubator (see Note 23). For in vitro fertilization, oocytes are matured for 40–42 h; for oocyte pathogenetic activation, oocytes are matured for 42–44 h.

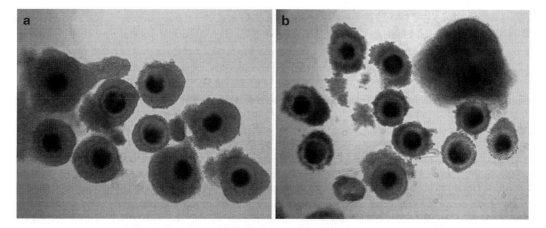

Fig. 3. Examples of selected (**a**) and nonselected (**b**) cumulus oocyte complexes isolated from abattoir derived porcine ovaries.

3.2. Day 2

1. Oocytes remain in maturation medium in the incubator on this day (see Note 24).

2. Warm HEPES and prepare medium for further manipulations, if necessary.

3.3. Day 3

3.3.1. Denuding Oocytes

1. Cumulus cells can be removed from a large number of oocytes using a vortex mixer (see Note 25). Place 100 μL SOF-HEPES into a 1.5 mL centrifuge tube and add 1 μL hyaluronidase stock (final concentration 100 μg/mL, 80–160 U/mL) and keep it on the slide warmer. Alternatively, 500 μL SOF-HEPES may be placed in to a 15-mL centrifuge tube and 5 μL hyaluronidase stock added. Keep individual treatments separate.

2. Oocytes are removed from maturation medium using the 5 μL handheld positive displacement pipette with glass bore tip (Drummond), and added directly to the tubes containing hyaluronidase in a minimal amount of maturation medium. Vortex the tube for 3 min on a medium-high setting. Microcentrifuge tubes are then rinsed three times with 1 mL SOF-HEPES using a 1 mL pipetman to remove the oocytes and 15-mL tubes are rinsed once with 5 mL SOF-HEPES and swirling. Rinses remove both oocytes and cumulus cells, which are now in suspension, from the vortex tubes.

3. Place the recovered SOF-HEPES with oocytes into an empty 35 × 10 mm dish, again using the positive displacement pipette. Oocytes are then moved to a new 35 × 10 mm dish with clean SOF-HEPES to wash (by swirling the dish). It is important to remove all hyaluronidase and separate the oocytes from the cumulus cells. Denuded oocytes are now ready for subsequent manipulation, such as in vitro fertilization, activation, or enucleation for nuclear transfer.

*3.3.2. Assessment
of Meiotic Stage*

1. After oocytes are denuded, they may be mounted and fixed to determine meiotic stage. Oocytes (approximately 10 per slide) are placed onto a glass microscope slide (Rite-On Micro Slides, Gold Seal) in <5 µL of SOF-HEPES medium with the pipette. As much medium as possible is then removed from the slide back into the pipette bore, leaving the oocytes on the slide with only a thin film of medium.

2. A small amount of mountant is placed in the middle of each side of a coverslip (22×22 mm, Fisher Scientific) and quickly placed over the oocytes. Each piece of mountant is alternatively pressed until oocytes are flattened, but not ruptured. The coverslip is adhered to the glass slide using rubber cement on opposite corners, and the slide labeled with a pencil. After a 1 min drying time, the slide is placed into a coplin jar with fixative for at least 48 h at room temperature.

3. Slides are then removed from fixative, and a drop of acetoorcein stain is placed at the edge of one corner of the coverslip. The stain is drawn underneath the coverslip by placing the edge of a KimWipe against the edge of the coverslip on the corner opposite the stain to absorb the fixative. Oocytes are examined at 400× magnification on a phase-contrast or inverted microscope. This procedure may be done at any time during the maturation process, although it is typically done when oocytes are removed from maturation to assess the percentage of oocytes completing the first meiotic division (at the metaphase II stage with first polar body; Fig. 4) (see Note 26).

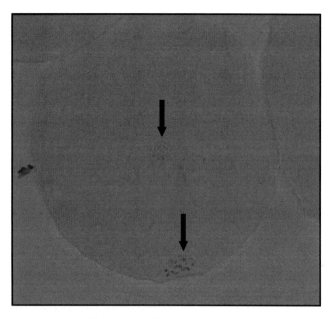

Fig. 4. A porcine oocyte fixed and stained with acetoorcein after maturation in vitro for 42 h, at the metaphase II stage of meiosis. As indicated by arrows, the metaphase plate can be seen in the middle of the ooplasm, and the polar body is apparent at the bottom of the oocyte.

4. Notes

1. When QS'ing and filtering SOF-HEPES with BSA, bubbles can be a problem. Let the BSA fully dissolve slowly with gentle stirring. Pour the solution slowly down the side of the volumetric flask.

2. All media is filtered with Millipore PVDF membranes, which are low protein binding. Each product has been tested for embryo toxicity. We do not discard the first 2 mL of medium filtered through the syringe filters when we make our working media, although this is common practice in some laboratories to ensure no toxins are washed from the filter.

3. Use TCM199 with Earle's salts, with L-glutamine, without sodium bicarbonate. Follow instructions on the package when preparing powdered TCM199.

4. Sterile, disposable, individually wrapped syringes used for filtering working solutions do not have a black rubber stopper with lubricant on the plunger, as the lubricant is embryotoxic. They are available in 5, 10 and 20 mL sizes. (Norm-ject, Henke, Sass, Wolf).

5. 100× PSA is composed of 10,000 IU/ML penicilin, 10,000 µg/ML streptomycin, and 25 µg/mL amphotericin.

6. LH and FSH stocks must be prepared, aliquoted into sterile 0.5 mL microfuge tubes in the amount needed for the maturation working solution once per month, as they can lose activity over time. Aliquots are stored at −20°C.

7. LH and FSH are sometimes substituted by 10 IU/mL equine chorionic gonadotropin (eCG) or pregnant mare serum gonadotropin (PMSG), and 10 IU/mL human chorionic gonadotropin (hCG) (27–29).

8. To prepare porcine follicular fluid for use in maturation medium, save the follicular fluid collected after maturation in the 50 mL conical centrifuge tubes after aspiration. Centrifuge these tubes at $3,000 \times g$ for 5 min. Take the supernatant and filter (0.8 µm, 25 mm Acrodisk, Pall Life Sci.). Make 1.2 mL aliquots in sterile 1.5-mL microfuge tubes, label, and store at −20°C.

9. Because TCM199 also contains 5.56 mM glucose, the final concentration in this maturation medium is 5.56 mM plus 3.05 mM, to give 8.61 mM.

10. 10% pFF can be replaced by 0.1% PVA to make defined NCSU23 maturation medium.

11. Fixative should be made fresh for oocyte fixation for best results.

12. Acetoorcein stain is difficult to filter because of the debris in the solution. Filter only 1–2 mL at a time.

13. All media must equilibrate for a minimum of 4 h before use in the incubator to ensure correct pH of approximately 7.35.

14. There are many kinds of sterile 60×15 and 35×10 culture dishes which may be used. The dishes we reference are non tissue culture treated. We use these dishes to create a smaller diameter, taller/deeper drop. Often, tissue culture treated dishes result in spreading of the medium drop resulting in a short, flat drop. Dishes specifically created and tested for IVF are also available and are of very high quality, although they are expensive.

15. There are multiple types of mineral oil available for oocyte maturation and embryo culture, of a variety of quality and cost. We use embryo tested oil and then wash it prior to use with PBS (200 mL PBS to 500 mL oil, shake and let separate) to remove toxic substances. We then place oil in the incubator to equilibrate before the preparation of maturation dishes.

16. TCM199 with HEPES buffer has also been used for ovary transport, and is thought by some to enhance oocyte quality (30, 31).

17. Although warm saline is added to the ovaries at the abattoir, the temperature commonly decreases during transport, dependent upon transport container and ambient temperature. We typically use thermos coolers to transport ovaries, which provide some insulation. If the outside temperatures are very cold, we place the thermos in a Styrofoam cooler containing warm packs.

18. Oocytes can also be isolated from ovaries by cutting the ovary. This works best for smaller (peripubertal gilt) ovaries. Each ovary is sliced in half along the long axis of the ovary and placed into a large petri dish containing SOF-HEPES. Follicles are punctured using a scalpel blade, and each half of the ovary is sliced in a checkerboard pattern. This is repeated on each side. The half ovary is then held with forceps and swirled in the SOF-HEPES to wash away the oocytes, and then discarded. Approximately four ovaries can be collected per dish. The HEPES medium is then poured from the petri dish in to a 50-mL conical tube and oocytes allowed to settle to the bottom. Care must be taken not to select small oocytes that are not fully grown, likely isolated from very small follicles, when using this method.

19. Time must be carefully observed when searching for, selecting and placing oocytes into maturation medium. The faster the oocytes can be processed and placed into the incubator, the better the oocyte quality will be.

20. Because oocytes are sensitive to light, we dim the overhead lights in our laboratory such that there are no lights directly

over the workspace. In addition, we place red theater film over the base of our stereoscopes to filter out harmful wavelengths while still allowing us to see the oocyte.

21. Drummond tips (bores) are sterilized by autoclaving prior to use, and should be changed when moving between dishes. Other methods of pipetting oocytes are also available. Unipettes can be pulled to the correct size over a flame and attached to a 1 mL syringe. The disadvantage of this method is that it is not positive displacement, so it is possible to lose oocytes within the pipette. However, smaller volumes can be used relative to the Drummond pipette. In addition, long Pasteur pipettes can be pulled to the appropriate size over a flame, and attached via a small piece of Tygon tubing to a 0.22-μm filter for protection, which has small diameter tubing attached on the opposing side that is then attached to a mouthpiece for mouth pipetting. It is also quite easy to lose oocytes in this type of pipette, but again smaller volumes are possible.

22. It is helpful during the selection process if horizontal lines (approximately 4 mm apart) are drawn on the bottom of the 60×15 mm selection dish prior to filling it with HEPES. During selection, oocytes can be individually examined while moving along each section.

23. We use Thermo/Forma 3110 culture incubators, although there are many good incubators available. The incubator should be water jacketed to maintain temperature, be able to regulate CO_2 (infra-red CO_2 monitors have a shorter recovery time, but also a shorter working lifespan). We do not use a dual gas incubator (capable of regulating both CO_2 as well as O_2 via nitrogen injection) for maturation, as we have found that reduced oxygen environments are detrimental to porcine oocyte maturation. The incubator water pan should be maintained at the fullest level possible to maintain 100% constant humidity. The incubator should be checked for temperature and CO_2 atmosphere each day with a calibrated thermometer and Fyrite sensor. CO_2 settings drift over time, so the incubator must be recalibrated 1–2 times per year. When going in and out of the incubator to retrieve and replace dishes, the door should not be completely opened and the time of opening should be minimized to avoid prolonged recovery time for the gas environment and temperature. In addition, whenever maturation medium dishes are outside the incubator, work must be completed very quickly because in the ambient environment, the pH of the medium as well as the temperature will change rapidly, resulting in a suboptimal environment and reduced oocyte quality.

24. Some protocols move oocytes from maturation medium containing hormones to hormone free medium at 22 h post-maturation for the remaining maturation period, particularly if 10 IU/mL PMSG/eCG and hCG are used in maturation medium (see Note 8) (28). We have found no difference in development whether oocytes are moved to hormone free medium halfway through the IVM period when using FSH/LH, and feel that it is more detrimental to remove the oocytes from the incubator.

25. Another method of removing cumulus cells is "stripping" using a small bore pipette. This is advantageous as it is less stressful on the oocytes, and does not risk disrupting the positioning of the polar body relative to the metaphase spindle for nuclear transfer purposes. In this method, prepare an organ culture dish (BD Falcon 3037) by adding 4 mL HEPES to the outer moat and 1 mL to the inner well. Add 10 µL hyaluronidase stock to the inner well. Place the oocytes to be denuded in the inner well. Initial denuding can be accomplished by pipetting oocytes up and down using a 200 µl pipettor set to ~100 µL. Each oocyte is then taken in and out of pulled glass pipettes or commercially available "stripper tips" of decreasing sizes, with the final tip being exactly the size of the oocyte including the zona pellucida. Once oocytes are free of cumulus cells, place them in the outer well to minimize hyaluronidase exposure. Do only as many oocytes at one time as can be processed quickly.

26. Pig oocytes typically reach the metaphase II (MII) stage with extrusion of the polar body sometime between 36 and 42 h postmaturation. It is typically for 85% of the oocytes placed into our IVM system to successfully complete nuclear maturation and reach MII. In our system, we see the germinal vesicle stage until approximately 18–24 h postmaturation (hpm), at which time the germinal vesicle undergoes breakdown (GVBD; approximately 24 hpm). Condensed chromatin (CC) is present at approximately 24–30 hpm; and metaphase I (MI) at approximately 30–36 hpm. Anaphase (A) and telophase (T) are only present for a short time and can occasionally be seen at approximately 36 hpm, respectively.

References

1. Brackett BG, Bousquet D, Boice ML et al (1982) Normal development following in vitro fertilization in the cow. Biol Reprod 27:147–158.

2. Eyestone WH, First NL(1989) Co-culture of early cattle embryos to the blastocyst stage with oviducal tissue or in conditioned medium. J Reprod and Fert 85:715–720.

3. Yoshida M, Mizoguchi Y, Ishigaki K et al (1993) Birth of piglets derived from in vitro fertilization of pig oocytes matured in vitro. Theriogenology 39:1303–1311.

4. Thompson JG, Gardner DK, Pugh PA et al 1995) Lamb birth weight is affected by culture system utilized during in vitro pre-elongation development of ovine embryos. Biol Reprod 53:1385–1391.

5. Machaty Z, Day BN, Prather RS (1998) Development of early porcine embryos in vitro and in vivo. Biol Reprod 59:451–455.

6. Kikuchi K, Kashiwazaki N, Noguchi J et al (1999) Developmental competence, after transfer to recipients, of porcine oocytes matured, fertilized and cultured in vitro. Biol Reprod 60:336–340.

7. Bavister BD (1995) Culture of preimplantation embryos: facts and artifacts. Hum Reprod Update 1:91–148.

8. Cohen J, Mandelbaum J, Planchot M (1980) [Maturation and in vitro fertilisation of human oocytes recovered after spontaneous ovulation or after induction of ovulation by use of gonadotrophins (preliminary results)]. Journal de Gynecologie, Obstetrique et Biologie de la Reproduction 9:523–530.

9. Edwards RG (1965) Maturation in vitro of human ovarian oocytes. Lancet 2:926–929.

10. Edwards RG, Bavister BD, Steptoe PC (1969) Early stages of fertilization in vitro of human oocytes matured in vitro. Nature 221:632–635.

11. Gandhi AP, Lane M, Gardner DK et al (2000) A single medium supports development of bovine embryos throughout maturation, fertilization and culture. Hum Reprod 15:395–401.

12. Tervit HR, Whittingham DG, Rowson LEA (1972) Successful culture in vitro of sheep and cattle ova. J Reprod Fert 30:493–497.

13. Grupen CG, Nagashima H, Nottle MB (1997) Role of epidermal growth factor and insulin-like growth factor-I on porcine oocyte maturation and embryonic development in vitro. Reprod Fert Dev 9:571–575.

14. Abeydeera LR, Wang WH, Cantley TC et al (1998) Presence of epidermal growth factor during in vitro maturation of pig oocytes and embryo culture can modulate blastocyst development after in vitro fertilization. Mol Reprod Dev 51:395–401.

15. Reed ML, Estrada JL, Illera MJ et al (1993) Effects of epidermal growth factor, insulin-like growth factor-I, and dialyzed porcine follicular fluid on porcine oocyte maturation in vitro. J Exper Zool 266:74–78.

16. Yoshida M, Ishizaki Y, Kawagishi H et al (1992) Effects of pig follicular fluid on maturation of pig oocytes in vitro and on their subsequent fertilizing and developmental capacity in vitro. J Reprod Fertil 95:481–488.

17. Herrick JR, Brad AM, Krisher RL (1998) Chemical Manipulation of Glucose Metabolism in Porcine Oocytes: Effects on Nuclear and Cytoplasmic Maturation. In Vitro Reprod 131:289–298.

18. Krisher RL, Brad AM, Herrick JR et al (2007) A comparative analysis of metabolism and viability in porcine oocytes during in vitro maturation. Anim Reprod Sci 98:72–96.

19. Sato H, Iwata H, Hayashi T et al (2007) The effect of glucose on the progression of the nuclear maturation of pig oocytes. Anim Reprod Sci 99:99–305.

20. Funahashi H, Kim NH, Stumpf TT et al (1996) Presence of organic osmolytes in maturation medium enhances cytoplasmic maturation of porcine oocytes. Biol Reprod 54:1412–1419.

21. Grupen CG, Nagashima H, Nottle MB (1995) Cysteamine enhances in vitro development of porcine oocytes matured and fertilized in vitro. Biol Reprod 53:173–178.

22. Yoshida M, Ishigaki K, Nagai T et al (1993) Glutathione concentration during maturation and after fertilization in pig oocytes: relevance to the ability of oocytes to form male pronucleus. Biol Reprod 49:89–94.

23. Yoshida M, Ishigaki K, Pursel VG (1992) Effect of maturation media on male pronucleas formation in pig oocytes matured in vitro. Mol Reprod Dev 31:68–71.

24. Petters RM, Wells KD (1993) Culture of pig embryos. J Reprod Fert Suppl 48:61–73.

25. Marques MG, Nicacio AC, de Oliveira VP et al (2007) In vitro maturation of pig oocytes with different media, hormone and meiosis inhibitors. Anim Reprod Sci 97:375–381.

26. Marchal R, Vigneron C, Perreau C, et al (2002) Effect of follicle size on meiotic and developmentla competence of porcine oocytes. Theriogenology 57:1523–1532.

27. Kishida R, Lee ES, Fukui Y (2004) In vitro maturation of porcine oocytes using a defined medium and developmental capacity after intracytoplasmic sperm injection. Theriogenology 62:1663–1676.

28. Funahashi H, Day BN (1993) Effects of the duration of exposure to hormone supplements on cytoplasmic maturation of pig oocytes in vitro. J Reprod Fert 98:179–185.

29. Funahashi H, Cantley T, Day BN (1994) Different hormonal requirements of pig oocyte cumulus complexes during maturation in vitro. J Reprod Fert 101:159–165.

30. Costa SHF, Andrade ER, Silva JRV et al (2005) Preservation of goat preantral follicles enclosed in ovarian tissue in saline or TCM 199 solutions. Small ruminant research: J Internat Goat Assoc 58:189–193.

31. Figueiredo JR, Hulshof SCJ, Hurk RVD et al (1993) Development of a combined new mechanical and enzymatic method for the isolation of intact preantral follicles from fetal, calf and adult bovine ovaries. Theriogenology 40:789–799.

Chapter 15

Experimental Approaches to the Study of Human Primordial Germ Cells

Andrew J. Childs and Richard A. Anderson

Abstract

The survival, proliferation, and differentiation of primordial germ cells in the mammalian embryo is regulated by a complex cocktail of growth factors and interactions with surrounding somatic cells, which together form a microenvironment known as the germ cell niche. Extensive insight into the signalling pathways that regulate PGC behaviour has been provided by the study of these cells in rodent models, however little is known about the factors that regulate these processes in human PGCs. In this review, we outline experimental approaches to the culture and manipulation of the first trimester human fetal ovary, and discuss immunohistochemical and stereological approaches to detect changes in human PGC numbers and proliferation in response to treatment with exogenous growth factors.

Key words: Gonad, Primordial germ cell, Ovary, Oocyte

1. Introduction

In human females, the entire oocyte population is believed to be established during fetal life, with no new oocytes generated postnatally. Understanding the interactions that occur between germ and somatic cells in the fetal ovary to regulate this process is therefore critical to our understanding of the female fertile lifespan and pathologies that curtail this (e.g. premature ovarian failure). Extensive research in murine models has identified many of the developmental stages and signalling pathways involved in regulating germ cell survival, proliferation and differentiation during fetal and postnatal life (1); however, the results of these studies cannot always be extrapolated to the human, as exemplified by differences in the onset of meiosis (discussed below) and sequential changes in the expression of germ cell markers such as the cytoplasmic RNA-binding proteins Dazl and Mvh (Mouse Vasa Homologue, also

Wai-Yee Chan and Le Ann Blomberg (eds.), *Germline Development: Methods and Protocols*, Methods in Molecular Biology, vol. 825, DOI 10.1007/978-1-61779-436-0_15, © Springer Science+Business Media, LLC 2012

known as Ddx4) (2). Elucidating the molecular mechanisms that underpin human gametogenesis may also generate new targets for contraceptives, and will help inform strategies to derive germ cells from pluripotent stem cells.

A summary of the key developmental events during human female fetal germ cell development is depicted in Fig. 1a. The precise timing of germ-line specification in the human embryo is unclear, but occurs prior to 4 weeks gestation, at which point migratory primordial germ cells (PGCs) are detectable on the hindgut (3). Colonization of the nascent gonad occurs from week 5 onwards, and sex determination shortly after (4, 5). PGC proliferation occurs throughout the migratory and colonization phases, and continues for several weeks after sex determination (6). The first germ cells in the human fetal ovary to enter meiosis do so around 11 weeks gestation (6–8). However, in contrast to the mouse, in which the onset of meiosis and germ cell differentiation occurs in a synchronized rostro-caudal wave along the axis of the gonad from embryonic day e13.5 onwards (9, 10), germ cell differentiation in the human fetal ovary is highly asynchronous, such that by the mid to late second trimester and beyond, proliferative PGC-like cells are still detectable alongside meiotic oogonia and oocytes that are associating with surrounding somatic cells to form the first primordial follicles (11–13).

The survival, proliferation, and differentiation of PGCs in the mammalian embryo is regulated by a complex cocktail of growth factors and interactions with surrounding somatic cells which together form a microenvironment known as the germ cell niche. Extensive insight into the signalling pathways that regulate PGC behaviour has been provided by the study of these cells in rodent models, however little is known about the factors that regulate these processes in human PGCs. Isolation and culture of murine PGCs on layers of feeder cells has provided an effective experimental system in which to identify growth factors which promote the survival or proliferation of murine primordial germ cells in culture (reviewed in refs. 14, 15). However, the effects of growth factors on isolated mouse PGCs have not always been replicated when the same signalling pathways have been disrupted in vivo (for discussion see refs. 16, 17). This approach has not been widely used to study the response of human PGCs to growth factors, and attempts to culture isolated human PGCs in vitro long term have focused instead on attempting to derive pluripotent Embryonic Germ cells (an Embryonic Stem cell-like cell derived from PGCs) (18, 19). Our laboratory and others have instead opted to use an organ culture approach to study the response of human PGCs' exogenous signals such as growth factors (8, 16, 20); reasoning that more physiologically representative responses to growth factors may be detected if a germ cell is maintained within its niche (i.e. with its associations with the surrounding gonadal cells retained intact)

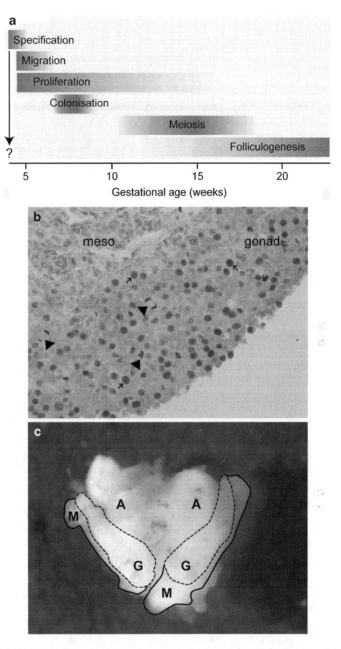

Fig. 1. (**a**) Timeline of key development events during human female fetal germ cell development. (**b**) Photomicrograph of section through a 65 days gestational age human fetal ovary cultured for 10 days in serum-free medium, and stained with germ cell marker AP-2γ. Note the abundant PGCs (*arrows*) and the presence of proliferating germ cells (*arrowheads*; meso: mesonephros). (**c**) Photomicrograph of dissected gonad–mesonephros–adrenal complex isolated from a 58 days gestational age human foetus. *Dashed line* outlines gonadal tissue. (*G* gonad, *M* mesonephros, *A* adrenal.).

than those seen in isolated PGCs cultured on monolayers. We have demonstrated that the first trimester (8–9 weeks gestation, 6–8 weeks post fertilization) human fetal ovary is capable of supporting germ cell survival and proliferation in serum-free media without exogenous growth factor support for up to 10 days (16), with excellent retention of tissue architecture (Fig. 1b). This suggests that the fetal ovary is capable of producing at least the minimal factors necessary to support germ cell survival. This is important, as a key challenge faced in working with human fetal ovaries is the relatively long doubling time of the germ cell population, estimated to be in the order of 6 days (6). Long-term culture approaches of greater than 1 week are therefore required to detect subtle effects on germ cell number or development.

Here we describe our approaches for the study of growth factor (e.g. BMP4) effects on germ cells in the first trimester human fetal ovary (16). The methods below detail the experimental approaches we have taken to detect changes in total germ cell number and proliferation using PGC marker Activator Protein-2gamma (AP-2γ; (21, 22)), and phosphorylated Histone H3 (12), across a long-term (10 days) culture period. The protocol also is adaptable for the study of changes in gene expression in response to growth factor treatment by reducing the culture period to 8–24 h, and collecting the tissues for RNA extraction rather than fixation for histology at the end of the culture. The method is suitable for ovaries from foetuses up to 10–11 weeks gestational age, as in our hands intact ovaries from foetuses older than this show increasing necrosis of non-peripheral regions of tissue during culture in vitro. Alternative approaches to the study of growth factor signalling in ovaries from later gestation specimens are detailed in (23–25).

2. Materials

2.1. Dissection, Tissue Collection and Cell and Organ Culture

All items should be sterilized (either by autoclaving (particularly for equipment/glassware) or by filter sterilization through a syringe filter with 0.2 μM pore size).

1. Serum-free culture media: MEMalpha (Invitrogen) supplemented with 1× MEM non-essential amino acids (diluted from 100× stock; Invitrogen), 2 mM sodium pyruvate (100 mM stock), 2 mM l-glutamine (200 mM stock), 1× Insulin, Transferrin and Selenium (ITS) supplement, (500× stock; Lonza), and 1× Penicillin/Streptomycin (100× stock), 3 mg/ml Bovine Serum Albumin (BSA).
2. Dulbecco's Phosphate Buffered Saline (DPBS).

3. PET membrane tissue culture inserts, 0.4 µM pore size (Greiner Bio-One), and 12 well tissue culture plates.

4. Recombinant human BMP4 (Invitrogen): reconstituted in 10 mM citric acid pH 3.0 and DPBS (containing 0.1% BSA) to a stock concentration of 10 ng/µl. Aliquot and store at −20°C.

5. Glass Petri dishes and cavity microscope slides.

6. Fine forceps, 28 G hypodermic needles, 1-ml disposable syringes.

7. Stereo dissecting microscope.

8. Bouin's Fixative.

2.2. Determination of Fetal Sex by PCR for SRY

1. Extraction Buffer: 25 mM sodium hydroxide, 0.2 mM ethylenediaminetetraacetic acid (EDTA).

2. Neutralization solution: 40 mM Tris.

3. TAE (Tris-acetate–EDTA) buffer: 0.04 M Tris-acetate, 0.001 M EDTA (dilute from 50× stock: 242 g Tris, 57.1 ml glacial acetic acid, 100 ml 0.5 M EDTA, pH 8.0, adjust to 1,000 ml with dH$_2$O.

4. ImmoMix Red (pre-mixed DNA polymerase/dNTPs/PCR buffer/gel loading dye; (Bioline).

5. DNA Oligonucleotide primers to *SRY*: Forward: 5′-ACAG TAAAGGCAACGTCCAG-3′, Reverse: 5′-ATCTGCGGGAA GCAAACTGC-3′ (26). Store 100 mM stocks at −20°C, and prepare 25 mM working stocks with ultrapure dH$_2$O.

6. 0.2 ml PCR tubes.

7. 1.5 ml microcentrifuge tube-compatible shaking heat block.

8. Programmable thermal cycler.

9. Agarose gel casting and electrophoresis tank (e.g. Mini-Sub Cell GT cell (Bio-Rad) or equivalent).

10. 2% TAE agarose gel: prepare by adding 1 g agarose (molecular/electrophoresis-grade), to 50 ml TAE buffer in a small Erlenmeyer flask and dissolve by heating in a microwave. Allow to cool slightly and add 5 µl GelRed Nucleic Acid Stain (Biotium (see Note 1)), mix well by swirling and pour into cast to set.

11. 100 basepair (bp) DNA ladder.

2.3. Immunohistochemical Detection of Germ Cells

1. 0.01 M citrate buffer: diluted from 0.1 M stock; 0.1 M citric acid in de-ionized dH$_2$O, adjusted to pH 6.0 with 5 M sodium hydroxide solution, store at 4°C.

2. Tris-buffered Saline (TBS): 0.01 M Tris, 0.85% sodium chloride, adjusted to pH 7.4 with concentrated hydrochloric acid.

3. 3% hydrogen peroxide: diluted in methanol from 30% (w/v) stock.

4. Blocking serum: TBS supplemented with 20% normal goat serum and 5% BSA (see Notes 2 and 3).

5. Primary antibody (e.g. rabbit polyclonal anti-AP-2γ; Santa Cruz Biotechnology): diluted 1:100 in blocking serum; (see Note 4) for detection of total germ cell numbers, or rabbit polyclonal anti-phosphoHistone H3 (phospho-H3; Upstate Biotechnology); diluted 1:100 in blocking serum for detection of proliferating cells). Store on ice until required.

6. Goat anti-rabbit biotinylated secondary antibody (DAKO): diluted 1:500 in blocking serum; see Note 3).

7. Horseradish peroxidase (HRP)-conjugated streptavidin (Vector Laboratories): diluted 1:1,000 in TBS.

2.4. Stereological Determination of Germ Cell Numbers

1. Stereo microscope (Axio Imager A1 (Carl Zeiss) or equivalent) with digital camera and automatic stage.

2. PC with stereology software (Image Pro Plus (Media Cybernetics) or equivalent).

3. Methods

3.1. Culture of Human First Trimester Fetal Ovaries

This method outlines our strategy to detect changes in germ cell number and proliferation as previously reported (16), but in principle, any growth factor or supplement could be substituted for the BMP4 used here. Owing to the wide variation in ethical attitudes towards – and regulatory frameworks governing – the use of human fetal tissues in research; a discussion of the ethical approval process, patient recruitment and tissue collection is outwith the scope of this article, but gonadal tissue can be obtained following both medical and surgical termination of pregnancy. We do not use material derived from spontaneous miscarriage.

1. Place the foetus in DPBS in a glass Petri dish. Remove a small piece of non-gonadal tissue (e.g. limb) for genotyping (see Subheading 3.2 below). Carefully puncture the skin over the liver, peel back, and remove the liver.

2. Remove the adrenal–mesonephros–gonad complex (Fig. 1c) from the abdomen by placing fine forceps underneath the structure and peeling it off. Place this into a drop of DPBS on a cavity slide on a dissection microscope, and sub-dissect the gonad–mesonephros complex away from the adrenals (see Note 5) using 28 G needles attached to 1 ml syringes. Transfer the gonad–mesonephros complexes to DPBS using fine forceps, manipulating the tissue by gripping the mesonephros only.

3. Prepare 1 ml of serum-free medium supplemented with 10 µl DPBS/BSA (control) and another 1 ml supplemented with 10 µl of human recombinant BMP4 (final concentration 100 ng/ml). Place 900 µl of each supplemented medium in adjacent wells of a 12 well plate, and place a tissue culture well insert into each (see Note 6) well. Place a drop (approximately 50 µl) of the corresponding media onto the upper surface of each membrane. Fill the remaining wells with 1 ml DPBS to humidify the plate.

4. Carefully transfer the gonad–mesonephros complexes into the drop of media on the upper surface of the membrane, and position in center of the well.

5. Cover plate, and culture in a humidified incubator at 37°C/5% CO_2. Perform a half media change (see Note 7) every 48 h.

6. After 7–10 days, remove the tissue culture inserts. Carefully pipette 200 µl of DPBS into the insert to wash the tissue.

7. Remove the gonad–mesonephros from the membrane and place in Bouin's Fixative (~4 ml per gonad–mesonephros) for 1 h at room temperature. Transfer to 70% ethanol for storage prior to processing into paraffin blocks.

3.2. Determination of Fetal Sex by PCR for SRY

The sex of fetal specimens is difficult to determine by gross fetal or gonadal morphology around 8–9 weeks gestation. To establish the sex of the foetus from which the tissue has been obtained (and thus determine whether the dissected gonad is an ovary or a testis), we take a small piece of non-gonadal tissue (e.g. the limb), extract DNA and perform PCR for the Y-linked gene *SRY*. The protocol detailed below (adapted from ref. 26) yields a ~300 base-pair product only in PCR reactions containing DNA from male foetuses. This procedure can be performed in around 3 h, during which time dissected gonads can be stored in serum-free media in a humidified incubator (37°C/5% CO_2).

1. Digest tissue in a 1.5-ml microcentrifuge tube containing 100 µl Extraction Buffer. Place at 95°C for 20 min in a heating block with vigorous shaking. Cool on ice for 5 min.

2. Add 100 µl Neutralization solution. Vortex vigorously for 30 s and place tube on ice.

3. Assemble PCR reactions as required in 0.2-ml PCR tubes. For each reaction, include 12.5 µl 2× ImmoMix Red, 0.5 µl 25 mM Forward *SRY* primer, 0.5 µl 25 mM Reverse *SRY* primer, 6.5 µl ultrapure dH_2O, and 5 µl template (extracted DNA solution subsequent to neutralization or dH_2O as a negative control; (see Note 8)).

4. Amplify *SRY* PCR amplicon on a programmable thermal cycler. Cycling conditions: 95°C for 10 min (hot start), followed by

35 cycles of 95°C for 30 s, 58°C for 30 s and 72°C for 45 s, and a final step of 72°C for 10 min.

5. Run 10 µl of PCR reactions on a 2% TAE agarose gel, alongside 10 µl 100-bp ladder. Determine presence of bands by UV transillumination; presence of a band indicates male specimen. Photograph gel as required.

3.3. Immunohistochemical Detection of Germ Cells

Following fixation, gonad–mesonephros complexes are processed into paraffin blocks by standard methods, and 8-µm sections are cut on a microtome (11). Single sections are mounted on consecutively numbered, electrostatically charged microscope slides before drying overnight in a slide oven. The mounting of adjacent sections on consecutive slides permits the total number of sections containing gonadal material to be established and thus enables the total countable gonadal area to be calculated as this is required for the calculation of germ cell number per gonad (27), if required. It also allows multiple different markers (such as antibodies to detect total, proliferating and apoptotic germ cells) to be performed on consecutive, comparable sections.

1. Dewax and rehydrate slides (xylene (2×5 min)), 100% ethanol (2×20 s), 95% ethanol (20 s), 70% ethanol (20 s), rinse with dH_2O.

2. Place slides in a pressure cooker containing 0.01 M (boiling) citrate buffer, pH 6.0, seal pressure vessel and boil for 5 min after the appearance of steam from the release valve. Release pressure and allow to cool for 20 min before retrieving slides. Cool in tap water.

3. Incubate slides for 30 min in 3% hydrogen peroxide in methanol to block peroxidase activity in the tissue.

4. Wash slides briefly 3× in dH_2O, followed by 2×5 min washes in TBS.

5. Transfer slides to a TBS-humidified slide tray. Incubate sections in blocking serum (approximately 100 µl per slide, or sufficient to cover the tissue sections entirely) for 30 min at room temperature.

6. Wash 3×5 min in TBS.

7. Taking each slide in turn, carefully dry around the edge of the tissue to remove excess TBS. Place one drop of streptavidin blocking solution (avidin-biotin blocking kit) on each tissue section. Incubate for 15 min at room temperature in a TBS-humidified slide tray.

8. Wash 3×5 min TBS.

9. Repeat step 7, applying instead, one drop of Biotin blocking solution; incubate for 15 min.

10. Wash for 3×5 min in TBS.

11. Apply primary antibody (e.g., anti-AP-2γ (see Note 9) or anti-phospho-H3 diluted in blocking serum, 50–100 μl per slide). Include negative controls (sections incubated in blocking serum minus primary antibody). Cover sections with small pieces of parafilm to prevent drying, and place in a humidified slide tray overnight at 1°C.

12. Wash 3×5 min in TBS.

13. Apply secondary antibody (50–100 μl per slide of goat anti-rabbit biotinylated antibody diluted in blocking serum) to sections, incubate in a humidified tray at room temp for 30 min.

14. Wash 3×5 min.

15. Incubate slides with HRP-conjugated streptavidin diluted in TBS, 50–100 μl per slide, in a humidified slide tray, for 30 min.

16. Wash 3×5 min.

17. For antibody detection, mix one drop of DAB chromogen with 1 ml diluent (DAB kit).

Take each slide in turn, dry off excess TBS and apply 50–100 μl DAB. Observe brown colour development under microscope. When appropriate staining intensity is reached, transfer slides in dH_2O to quench reaction and prevent further staining.

To provide contrast to the brown DAB staining, and to highlight tissue architecture, sections are then counterstained with haematoxylin (blue), before dehydration of sections through graded alcohols and xylene (Subheading 3.3 step 1 above, in reverse order, i.e. 70% ethanol to xylene). Sections can then be mounted permanently under glass cover slips using a mounting medium such as Pertex (CellPath Ltd, Newtown Powys, UK). Once dry, staining intensity and localization can be assessed by microscopy, or stereological assessment of germ cell number/proliferation performed as below.

3.4. Stereological Determination of Germ Cell Number

Stereology enables the quantification of the number of objects of interest (e.g. germ cells, proliferating cells), in a three dimensional tissue or organ from counting two dimensional tissue sections. Although numerous methods for this are available, we have used the method detailed in ref. 27 and found that counting every fifth or tenth section of a serially sectioned first trimester fetal ovary is sufficient to generate accurate cell numbers. In our studies, we determine both the total germ cell number (AP-2γ-positive cells) within an ovary, and the total proliferating (phospho-H3-positive) germ cell number in the same gonad, as this allows the latter to be expressed as a proportion (i.e. percentage of total germ cells that are proliferating). Changes in germ or somatic cell number may also alter the volume of the ovary; therefore, calculating the total countable ovarian area allows the number of germ cells per unit area to be calculated, which may be a more accurate measure than the total germ cell number within the ovary. Details of the methodology will vary according to software and microscopy equipment used. The protocol below is therefore an outline of the key procedures undertaken during stereological assessment of cell numbers in the fetal gonad.

1. Scan slide to produce a tiled image of the tissue section using a 40× objective lens.

2. Count stained cells by manually tagging each cell once within the gonadal area using a point-tag tool. The number of tags will be calculated automatically.

3. Using an outline tool, determine the area (in μm²) of gonadal tissue on the section by drawing a line around the perimeter of the area of interest.

4. If more than one area per slide is to be counted, total the individual counted cells and areas.

5. Calculate the cells per area by dividing the number of cells counted by the total area.

4. Notes

1. GelRed Nucleic Acid Stain is a non-toxic substitute for ethidium bromide for visualizing DNA fragments by UV transillumination. However, if required, ethidium bromide can be used instead at a final concentration of 0.5 μg/ml.

2. If a secondary antibody is used that is raised in a species other than goat, ensure that the normal goat serum used here is replaced with serum from the same species in which the secondary antibody is raised.

3. Blocking serum can be prepared in bulk and stored in appropriately sized aliquots (approximately 10 ml) at −20°C.

4. The concentrations of primary and secondary antibodies suggested here are guidelines based on our experience with these products. Batch to batch variation in concentration and affinity for the antigen means that the optimum concentration should always be determined empirically by testing several different concentrations of primary antibody in parallel.

5. Owing to the fragility of the gonad, we normally culture first trimester fetal ovaries attached to the mesonephros. This provides structural support to the tissue, but is also a way of manipulating the tissue without gripping the gonadal tissue directly. If it is thought that retaining the mesonephros may confound the effect(s) being investigated, the gonad can be

detached from the mesonephros by sub-dissection on a cavity slide with 28 G needles and cultured independently as above.

6. The membrane of the tissue culture well insert should sit on the meniscus of the culture media in the well, such that the gonad–mesonephros is cultured on an air–liquid interface when the membrane has equilibrated. The volumes given here work with the insert–plate combinations detailed in Subheading 2.1. Combinations of inserts/plates from other manufacturers may require optimization. Note that the tissue will become completely immersed if the volume of medium in the well is below the membrane is too great, leading to a loss of the air–liquid interface.

7. To perform a half media change, carefully remove 450 μl of media from the well outside the tissue culture insert using a P1000 pipette. Replace this with the same volume of fresh media.

8. Include suitable controls for the PCR reaction, such as one reaction containing ultrapure water in place of fetal DNA (no template, thus negative control), and one reaction containing template DNA from a specimen previously confirmed to be male (positive control).

9. For the detection of human PGCs, antibodies raised against OCT4 can be used in place of AP-2γ.

Acknowledgements

We are grateful to Hazel Kinnell, Dr Rosemary Bayne, Sharon Eddie, and the staff of the Medical Research Council Human Reproductive Sciences Unit Histology Core Facility for technical assistance in developing these protocols, and to Anne Saunderson, Joan Creiger, and the staff of the Bruntsfield Suite of the Royal Infirmary of Edinburgh for assistance with patient recruitment and specimen collection. This work is supported by Medical Research Council core funding to RAA (U.1276.00.002.00001.01) and a Medical Research Scotland grant (354 FRG) to AJC.

References

1. Matzuk MM, Lamb DJ (2008) The biology of infertility: research advances and clinical challenges. Nat Med 14:1197–1213.

2. Anderson RA, Fulton N, Cowan G, Coutts S, Saunders PT (2007) Conserved and divergent patterns of expression of DAZL, VASA and OCT4 in the germ cells of the human fetal ovary and testis. BMC Dev Biol 7:136.

3. Witschi E (1948) Migration of the germ cells of human embryos from the yolk sac to the primitive gonadal folds Contrib Embryol 32:67–80.

4. Hanley NA, Hagan DM, Clement-Jones M, Ball SG, Strachan T, Salas-Cortes L, McElreavey K, Lindsay S, Robson S, Bullen P, Ostrer H, Wilson DI (2000) SRY, SOX9, and DAX1

expression patterns during human sex determination and gonadal development. Mech Dev 91:403–407.

5. Wartenburg H (1981) Differentiation and development of the testes, in The Testis (Burger, H., de Kretser, D.M., Ed.), 2nd ed., Raven Press, New York.

6. Bendsen E, Byskov AG, Andersen CY, Westergaard LG (2006) Number of germ cells and somatic cells in human fatal ovaries during the first weeks after sex differentiation. Hum Reprod 21:30–35.

7. Gondos B, Westergaard L, Byskov AG (1986) Initiation of oogenesis in the human fatal ovary: ultrastructural and squash preparation study. Am J Obstet Gynecol 155:189–195.

8. Le Bouffant R, Guerquin MJ, Duquenne C, Frydman N, Coffigny H, Rouiller-Fabre V, Frydman R, Habert R, Livera G (2010) Meiosis initiation in the human ovary requires intrinsic retinoic acid synthesis. Hum Reprod 25:2579–2590.

9. Bullejos M, Koopman P (2004) Germ cells enter meiosis in a rostro-caudal wave during development of the mouse ovary. Mol Reprod Dev 68:422–428.

10. Menke DB, Koubova J, Page DC (2003) Sexual differentiation of germ cells in XX mouse gonads occurs in an anterior-to-posterior wave. Dev Biol 262:303–312.

11. Anderson G, Gordon K (1996) Tissue processing, microtomy and paraffin sections, in Theory and Practice of Histological Techniques (Bankroft, J. D., and Stevens, A., Eds.) 4th ed., Churchill Livingstone, New York.

12. Fulton N, Martins da Silva SJ, Bayne RA, Anderson RA (2005) Germ cell proliferation and apoptosis in the developing human ovary. J Clin Endocrinol Metab 90:4664–4670.

13. Stoop H, Honecker F, Cools M, de Krijger R, Bokemeyer C, Looijenga LH (2005) Differentiation and development of human female germ cells during prenatal gonadogenesis: an immunohistochemical study. Hum Reprod 20:1466–1476.

14. De Felici M, Scaldaferri ML, Lobascio M, Iona S, Nazzicone V, Klinger FG, Farini D (2004) Experimental approaches to the study of primordial germ cell lineage and proliferation. Hum Reprod Update 10:197–206.

15. Donovan PJ (1994) Growth factor regulation of mouse primordial germ cell development. Curr Top Dev Biol 29:189–225.

16. Childs AJ, Kinnell,HL, Collins CS, Hogg K, Bayne RA, Green SJ, McNeilly AS, Anderson RA (2010) BMP signaling in the human fatal ovary is developmentally regulated and promotes primordial germ cell apoptosis. Stem Cells 28:1368–1378.

17. Childs AJ, Saunders PT, Anderson RA (2008) Modelling germ cell development in vitro. Mol Hum Reprod 14:501–511.

18. Shamblott MJ, Axelman J, Wang S, Bugg EM, Littlefield JW, Donovan PJ, Blumenthal PD Huggins GR, Gearhart JD (1998) Derivation of pluripotent stem cells from cultured human primordial germ cell., Proc Natl Acad Sci USA 95:13726–13731.

19. Turnpenny L, Brickwood S, Spalluto CM, Piper K, Cameron IT, Wilson DI, Hanley NA (2003) Derivation of human embryonic germ cells: an alternative source of pluripotent stem cells. Stem Cells 21:598–609.

20. Lambrot R, Coffigny H, Pairault C, Donnadieu AC, Frydman R, Habert R, Rouiller-Fabre V (2006) Use of organ culture to study the human fatal testis development: effect of retinoic acid. J Clin Endocrinol Metab 91:2696–2703.

21. Hoei-Hansen CE, Nielsen JE, Almstrup K, Sonne SB, Graem N, Skakkebaek NE, Leffers H, Rajpert-De Meyts E (2004) Transcription factor AP-2gamma is a developmentally regulated marker of testicular carcinoma in situ and germ cell tumors. Clin Cancer Res 10:8521–8530.

22. Pauls K, Jager R, Weber S, Wardelmann E, Koch A, Buttner R, Schorle H (2005) Transcription factor AP-2gamma, a novel marker of gonocytes and seminomatous germ cell tumors. Int J Cancer 115:470–477.

23. Bayne RA, Eddie SL, Collins CS, Childs AJ, Jabbour HN, Anderson RA (2009) Prostaglandin E2 as a regulator of germ cells during ovarian development. J Clin Endocrinol Metab 94:4053–4060.

24. Childs AJ, Bayne RA, Murray AA, Martins Da Silva SJ, Collins CS, Spears N, Anderson RA (2010) Differential expression and regulation by activin of the neurotrophins BDNF and NT4 during human and mouse ovarian development. Dev Dyn 239:1211–1219.

25. Farhi J, Fisch B, Garor R, Peled Y, Pinkas H, Abir R (2010) Neurotrophin 4 enhances in vitro follicular assembly in human fatal ovaries, Fertil Steril. doi: doi:10.1016/j.fertnstert.2010.03.051.

26. Friel A, Houghton JA, Glennon M, Lavery R, Smith T, Nolan A, Maher M (2002) A preliminary report on the implication of RT-PCR detection of DAZ, RBMY1, USP9Y and Protamine-2 mRNA in testicular biopsy samples from azoospermic men. Int J Androl 25:59–64.

27. McClellan KA, Gosden R, Taketo T (2003) Continuous loss of oocytes throughout meiotic prophase in the normal mouse ovary. Dev Biol 258:334–348.

Chapter 16

Investigating the Origins of Somatic Cell Populations in the Perinatal Mouse Ovaries Using Genetic Lineage Tracing and Immunohistochemistry

Chang Liu, Melissa Paczkowski, Manal Othman, and Humphrey Hung-Chang Yao

Abstract

Genetic lineage tracing (or fate mapping) techniques are designed to permanently label progenitor cells of target tissues, thereby allowing delineation of the progenies of labeled cells during organogenesis. This technology has been widely used in the study of cell migration and lineage specification in various organs and organisms. Here, we describe how to apply the genetic lineage tracing model in combination with immunohistochemistry to identify the potential origins of somatic cell precursors in perinatal mouse ovaries.

Key words: Lineage tracing, Fate mapping, Immunohistochemistry, Progenitor, Somatic cells, Ovary, Organogenesis

1. Introduction

Traditional lineage tracing experiments utilize mostly fluorescent dyes to stain living cells, which pass the dye onto their progenies. One concern of this approach is that the dyes could diffuse to adjacent but unrelated cells (1). The Cre recombinase-mediated genetic lineage tracing technique is a reliable method that marks particular cell populations and their descendants during development (1, 2). The power of mouse genetics has allowed us to trace the fate of certain cells in not only a cell type-specific, but also a time-course dependent, manner. One of such systems requires a mouse line that carries the tamoxifen-inducible Cre recombinase gene under the control of a cell type-specific promoter. Treatment of tamoxifen activates the Cre recombinase exclusively in the cells that express the cell type-specific promoter. When a Cre reporter gene,

Wai-Yee Chan and Le Ann Blomberg (eds.), *Germline Development: Methods and Protocols*, Methods in Molecular Biology, vol. 825, DOI 10.1007/978-1-61779-436-0_16, © Springer Science+Business Media, LLC 2012

such as *Rosa26YFP* or *Rosa26LacZ*, is also present, the induced Cre recombinase turns on the reporter gene permanently in the target cells, therefore marking the cells and their progenies. Cre recombinase is active only in the presence of tamoxifen, allowing cell labeling to occur only in a defined period of time (3–6). This inducible lineage tracing model is different from traditional reporter lines in which the expression of the reporter is under the control of the promoter elements of the cell type-specific gene (7). The reporter in the traditional model reflects the endogenous expression pattern of the cell type-specific gene. If the gene is turned off during development, the reporter expression is off, consequently resulting in the inability to track the progenies of this particular cell type.

The inducible Cre-mediated lineage tracing technique has been successfully used in various organ systems, such as limb, brain, etc. (5, 8–10). However, such a genetic system has not been applied to the ovary, mainly because we have limited knowledge on genes implicated in establishing cell lineages in the ovary. We have been searching for the potential progenitor cell source of the theca cells, the mesenchymal cell types that produce androgens during folliculogenesis. It was found that the mesenchyme surrounding the follicles in the adult mouse ovary is positive for intracellular components of the Hedgehog signaling pathway, such as *Gli1* transcription factor (11–13). During fetal life, *Gli1* expression is absent in the ovary but is positive in the neighboring mesonephros, which contains the Wolffian and Müllerian ducts. After birth, *Gli1*-expressing cells first appear in the junction between the ovary and mesonephros and extend their presence to the mesenchyme of the neonatal ovary (unpublished data). We, therefore, hypothesize that *Gli1* can be used as a lineage tracing marker to investigate whether cells in the fetal mesonephros contribute to mesenchymal cell population in the neonatal ovary.

The lineage tracing experiments were accomplished by crossing the *Gli1^{tm3(cre/ERT2)Alj}*/J (or *Gli1-CreER^{T2}* hereafter) to the Cre reporter line (*Rosa26YFP* or *Rosa26LacZ*) (8, 14, 15). *Gli1-CreER^{T2}* mice express the tamoxifen-inducible Cre recombinase (CreER^{T2}) from the endogenous *Gli1* locus (8, 16). The *Rosa26YFP* mice contain an enhanced yellow fluorescent protein (*EYFP*) gene inserted into the ubiquitous *Rosa26* locus (Fig. 1). The expression of *EYFP* is blocked by an upstream *loxP*-flanked STOP sequence (15, 17). Tamoxifen treatment activates the Gli1-CreER^{T2}, which in turn excises the STOP sequence and allows the permanent expression of *EYFP* in the *Gli1*-positive cells and their progenies (Fig. 1). By giving tamoxifen during fetal stages when *Gli1* expression is present in the mesonephros but absent in the ovary, we can investigate whether the *Gli1*-positive cells in the mesonephros eventually become a part of the neonatal ovaries.

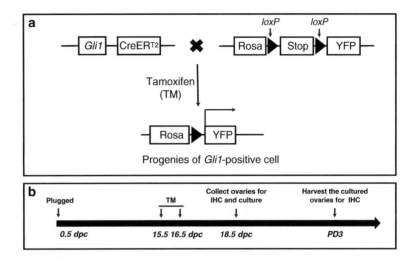

Fig. 1. (**a**) The lineage tracing strategy that marks and tracks *Gli1*-expressing cells. *CreER^T2* is under the control of the *Gli1* promoter elements and therefore expressed only in the *Gli1*-positive cells. In the presence of tamoxifen (TM), *CreER^T2* is activated and able to remove the *loxP*-flanked STOP sequence from the *Rosa26* allele, allowing a permanent *YFP* expression in the *Gli1*-expressing cells and their progenies only during the period of TM treatment. (**b**) Timeline of the experiment. The pregnant female is treated with TM at 15.5 and 16.5 dpc. At 18.5 dpc, the embryos are isolated and their ovaries were collected. For each embryo, one ovary is fixed for immunohistochemistry (IHC) and the other is cultured for 72 h until postnatal day 3 (PD3). After 3 days of culture, ovaries are harvested for immunohistochemistry.

2. Materials

2.1. Strains of Mice

Mice were maintained on a C57BL/6 J background and were used for the experiment at 3–5 months old for female and 3–8 months old for male.

1. *Gli1tm3(cre/ERT2)Alj/*J mice (8) (Heterozygote; Stock # 007913; The Jackson Laboratories).

2. B6.129X1-*Gt(ROSA)26Sortm1(EYFP)Cos/*J mice (15) (Homozygote; Stock # 006148; The Jackson Laboratories).

3. B6.129S4-*Gt(ROSA)26Sortm1Sor/*Jmice(14,18)(Homozygote; Stock # 003474; The Jackson Laboratories).

2.2. Tamoxifen Stock Solution

1. 70% ethanol: Add 30 mL of sterile water to 70 mL of 100% ethanol.

2. 50 mg/mL tamoxifen solution: Dissolve 0.05 g of tamoxifen in a solution of 950 µL corn oil and 50 µL 100% ethanol at 55°C in water bath. Vortex every few minutes until the powder is completely dissolved (see Note 1).

2.3. Genotyping

1. 50 mM NaOH solution.

2. 1 M Tris–HCl: Dissolve 121 g Tris base in sterile water. Adjust pH to 8.0 by adding 42 mL of HCl. Bring volume up to 1 L. Autoclave for 30 min.

3. PCR primers:

   Cre.465.633: 5′-GAG TGA ACG AAC CTG GTC GAA ATC AGT GCG-3′.
   Cre.58.87: 5′-GCA TTA CCG GTC GAT GCA ACG AGT GAT GAG-3′.

4. Mango PCR Mix (BioLine).

5. 50× TAE buffer: Dissolve 242 g Tris base in 57.1 mL glacial acetic acid, and 100 mL 0.5 M EDTA; add 750 mL distilled H_2O (dH_2O) and adjust pH to 8.0. Bring the final volume of solution to 1 L with dH_2O and autoclave for 30 min.

6. 0.5 M EDTA (pH 8.0): Add 186.1 g EDTA to 800 mL of double-distilled H_2O. Adjust pH to 8.0 and bring final volume to 1 L. Autoclave for 30 min.

7. 4% paraformaldehyde in 1× PBS.

8. 1.5% agarose gel (100 mL gel): In a small beaker, add 1.5 g agarose to 100 mL 1× TAE (2 mL 50× TAE with 98 mL ddH_2O). Heat the solution to boiling in the microwave to dissolve the agarose until no residues are seen in the solution. Let the solution cool for about 2 min and add 5 μl of ethidium bromide (0.005%) to the solution and mix. Pour the agarose solution into a gel box mold and let the gel cool to room temperature.

2.4. Organ Culture

1. 4-well culture plate.

2. 10× PBS: Weigh 80 g NaCl, 2 g KCl, 14.4 g NaH_2PO_4, and 2.4 g KH_2PO_4; add 800 mL of ddH_2O and pH to 7.4; adjust the final volume to 1 L and autoclave for 30 min.

3. Dissecting PBS: 1× PBS with 5–10% fetal bovine serum (FBS).

4. Bottle-top filters (150 mL cellulose acetate, 0.4 μm, 35-mm neck).

5. Culture medium: 1 mg/mL BSA, 1 mg/mL Albumax, 0.05 mg/mL ascorbic acid, 0.0275 mg/mL transferrin, 0.005 mg/mL streptomycin in Ham's F-12/DMEM.

   (a) In 50 mL Ham's F-12/DMEM, add 250 μL of 200 mg/mL BSA, 500 μL of 100 mg/mL Albumax (Invitrogen), 50 μL of 50 mg/mL ascorbic acid, 50 μL of 27.5 mg/mL transferrin (Sigma–Aldrich), 50 μL of 5,000 μg/mL streptomycin (1,000×) (Invitrogen).

(b) Filter the media through a bottle-top filter. Aliquot the media in sterile 50-mL tubes and store at 4°C. The medium is good for at least 2 weeks.

6. Millicell culture plate inserts (0.4 μm, 35-mm diameter).

2.5. Immunohisto-chemistry

1. 1× PBS.

2. 10, 15, and 20% sucrose in 1× PBS.

3. Optimum cutting temperature (OCT) compound.

4. 20% sucrose and OCT mix (1:1 and 1:3; v/v).

5. Blocking buffer: Combine 2.5 mL donkey serum (final concentration 5%; Jackson's Immunoresearch Laboratories), 50 μL of Triton X-100 (final concentration 0.1%), and 47.5 mL PBS.

6. Washing solution: Combine 0.5 mL donkey serum (final concentration 1%), 50 μL of Triton X-100 (final concentration 0.1%), and 49.5 mL PBS.

7. VECTASHIELD® Mounting Medium with DAPI.

8. Superfrost Plus Microscope Slides.

9. PAP/ImmEdge pen.

10. Microtome Cryostat.

11. Primary and secondary antibodies.

(a) Primary antibody: Goat anti-FOXL2 antibody (Imgenex Corp).

(b) Secondary antibody: Cy3-conjugated Donkey Anti-Goat IgG (Jackson ImmunoResearch Laboratories).

3. Methods

3.1. Breeding of Animals and Tamoxifen Treatment

1. Set up breeding cages for male *Gli1-CreER*$_{T2}$ and female *Rosa26YFP* or female *Rosa26LacZ* mice in the late afternoon (see Note 2).

2. The following morning, check each female for the presence of a vaginal copulation plug. If a copulation plug is observed, noon of that day is considered 0.5 days post coitum (dpc). Noon of the following day is considered 1.5 dpc and so forth.

3. On 15.5 dpc, the tamoxifen solution is fed orally to the pregnant female at 9 a.m. in the morning. Tamoxifen is given once per day for 2 consecutive days based on the body weight of the pregnant female (20 μL of the 50 mg/mL of tamoxifen stock solution per 10 g of body weight) (see Notes 3 and 4).

4. Euthanize the pregnant female at 18.5 dpc (following the IACUC guidelines) and dissect out the ovaries from the embryos. For each embryo, one ovary is used for organ culture (see Subheading 3.2) and the other is fixed in 4% paraformaldehyde at 4°C overnight (see Subheading 3.4 and Note 5). Collect a small piece of tissue (tail or limb, ~0.5 cm) from each embryo for genotyping and transfer the tissue to a 1.5-mL microcentrifuge tube. Store the tissue pieces at –20°C if they cannot be genotyped immediately (see Subheading 3.3 and Note 6).

3.2. Organ Culture (see Note 5)

3.2.1. Setting Up the Floating Filter Culture System

1. Using sterile techniques, cut the membrane of a Millicell culture plate insert into six pieces using a razor or scalpel.

2. Pipette 500 μl room temperature culture medium into each well of 4-well culture plates.

3. Lay one piece of Millicell membrane onto the medium in each well and allow the membrane to float on the medium. Put the culture plates in the incubator (37°C with 5% CO_2/95% air) for at least 30 min before culture to allow for equilibration of the media.

3.2.2. Tissue Collection and Culture Procedure

1. Dissect out the ovary with the mesonephros attached (referred to as the ovary complex hereafter) in dissecting PBS and wash them in room temperature culture medium for 5 min in the tissue culture hood. Use a single-channel pipettor with 200–1,000-μl tip to transfer each ovary complex onto the membrane. Place a drop of medium from that well onto the membrane so that the drop (approximately 2× the size of the ovary complex) of the medium covers the entire ovary (see Note 7).

2. The ovary complexes are cultured in a humidified incubator at 37°C with 5% CO_2/95% air for 3 days. Change culture medium every 24 h by removing most of the medium and replacing with the same volume of prewarmed medium (see Note 8). After 3 days of culture, collect samples for immunohistochemistry (see Subheading 3.4).

3.3. Genotyping (see Note 9)

1. Add 400 μL of 50 mM NaOH to each tube that contains the tissue pieces from Subheading 3.1 and place the tubes on a heating block at 95°C for 15 min or until the tissue has been digested (see Note 10). Tubes can be left on the heating block for 1–2 h, but should not be left overnight.

2. Add 200 μL of Tris–HCL (pH 8.0) to each tube, vortex the tubes, and centrifuge for 5 min at $16,100 \times g$ at room temperature. This sample is referred as "Extracted DNA" thereafter (see Note 10). For PCR, dilute gene-specific primers to 20 μM in PCR clean water (see Note 11).

3. For each PCR reaction, combine the following ingredients in a PCR tube:

Mango PCR mix	12.5 µl
Forward primer	0.625 µl
Reverse primer	0.625 µl
H_2O	10.25 µl
Extracted DNA	1 µl

4. Perform PCR using the following protocol:

Step 1	94°C for 2 min
Step 2	94°C for 45 s 60°C for 45 s 72°C for 1 min Repeat step 2 for 40 times
Step 3	72°C for 2 min and then hold at 10°C

5. Run PCR samples (25 µL) in 1× TAE buffer on a 1.5% agarose gel to determine the size of the PCR products (*Gli1Cre/+*: 408 base pairs; *Gli1+/+*: no band; see Note 9).

3.4. Immunohisto-
chemistry
(see Note 12)

Day 1

1. Collect tissues (18.5 dpc ovary complexes and the complexes after a 72-h culture; see Subheading 3.2) and fix them in 4% paraformaldehyde overnight at 4°C (no longer than 16 h) or 1–2 h at room temperature.

Day 2

1. Pipette off paraformaldehyde from the tube and add 1 mL PBS at room temperature. Wash the ovaries twice with 1 mL PBS at room temperature in 1.5-mL microcentrifuge tubes for 10 min each. The samples can be stored in PBS at 4°C for a few weeks. For long-term storage, store in 100% ethanol at –20°C.

2. Dehydrate the samples as follows:

1 mL 10% sucrose/PBS for 15 min
1 mL 15% sucrose/PBS for 15 min
1 mL 20% sucrose/PBS for 1 h
1 mL 20% sucrose/OCT (1:1) at room temperature for 4 h or overnight at 4 C

3. Transfer samples to an embedding block and remove the sucrose/OCT as much as possible. Fill the embedding block with 1:3 20% sucrose/OCT and position the samples using a toothpick under a dissection scope if necessary. Freeze samples

on dry ice (the frozen block can be stored in the microcentrifuge tubes at –20 or –70°C).

4. Use Microtome Cryostat to section the frozen block to 8–10-μm thickness. Put the sections on Superfrost plus microscope slides and dry the sections at room temperature for 15–30 min. Slides can be stored at 70°C in a slide box if not used immediately.

5. Circle desired sections with a PAP/ImmEdge pen and let dry. Add 500–1,000 μl 1× PBS at room temperature onto the circled area containing sections three times for a 5-min each wash.

6. Add 500–1,000 μl blocking buffer to the PAP/ImmEdge pen circled area till all the sections are immersed in the buffer. Put the slides in a humidified chamber for 0.5–1 h at room temperature.

7. Remove the blocking buffer from the slides.

8. Apply 500–1,000 μl Goat anti-FOXL2 primary antibody solution (1:300 in the blocking buffer) onto the sections to make sure that all the sections are immersed with antibody solution. Incubate the slides in a humidified chamber overnight at 4°C.

9. Rinse the sections three times by adding 500–1,000 μl 1× PBS at room temperature onto the sections for a 5-min each wash.

10. Apply 500–1,000 μl donkey anti-Goat IgG secondary antibody solution (1:500 in the blocking buffer) onto the sections and incubate the slides in a humidified chamber at room temperature for 1 h.

11. Rinse the sections three times by adding 500–1,000 μl 1× PBS at room temperature onto the sections for a 5-min each wash.

12. Add one or two drops of the mounting medium (DAPI) onto the sections and cover the slides with coverslips. The volume of mounting medium should be able to immerse all the sections after covered with coverslips. The slides can be stored at –20°C

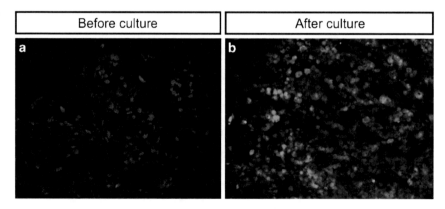

Fig. 2. Fluorescent immunohistochemistry for YFP-labeled *Gli1*-positive (*green*) cells and FOXL2-positive granulosa cells (*red*) in the ovary before culture (18.5 dpc) and after culture (PN3).

in the slide box for 2 weeks. Observe staining under a fluorescent microscope (Leica DMI 4000B) (see Fig. 2 and Note 12).

4. Notes

1. Tamoxifen solution made in corn oil is stored at 4°C for a week or at –20°C for several months. Warm the solution in 55°C water bath every time before use to ensure complete dissolution. Cool the dissolved solution to room temperature before dosing (16).

2. To get a higher rate of plugging, one male can be set up with two female mice at the same time. The copulation plug remains for only 12–14 h and then dissolves. Checking for the copulation plug early in the morning prevents the likelihood of missing any pregnancies.

3. It was shown that only a small percentage of cells were labeled by a single tamoxifen treatment. Multiple tamoxifen treatments provide a cumulative effect that mark more cells of interest (8).

4. Tamoxifen remains active for about 36 h after dosing. In this experiment, we dose the mouse at 15.5 and 16.5 dpc and collect ovaries at 18.5 dpc when tamoxifen is metabolized and excreted already (19–21).

5. Tamoxifen-treated dams often experience dystocia due to the inhibitory effects of tamoxifen on estrogen signaling. As a result of this complication, pups often die and cannot survive to adulthood. We, therefore, collect the ovaries before birth and put them in culture to study their development after birth. In addition to organ culture, one alternative is to perform cross fostering. Caesarian section is performed at 18.5 dpc and the pups are nursed by a foster dam. This allows the animals to survive to adulthood.

6. Because the genotypes of the embryos are not known at the time of dissection, the organ culture is set up blindly. Proper labeling of the cultured tissue (or culture dishes) is necessary to track the genotypes once the culture is completed.

7. You can use forceps to pick up a small amount of medium and then release the medium onto the membrane around the ovary. Alternatively, medium can be placed onto the ovary using a micropipette. One or two ovaries are cultured on each piece of membrane.

8. When removing medium, tilt the dish slightly and place pipette tip against the bottom edge of the well at the deepest part of the medium. When replacing medium to tissue-containing wells, pipette the medium with the tip against the wall of the

well so that the medium runs down the wall. Otherwise, the membrane may sink to the bottom.

9. For *Gli1CreER^{T2}* (heterozygous) and *Rosa26YFP* (homozygous) breeding, the possible genotypes of the pups are *Gli1^{Cre/+}*; *Rosa26^{YFP/+}* or *Gli1^{+/+}*; and *Rosa26^{YFP/+}*. The pups carrying the Cre allele are the experimental samples and Cre-negative samples serve as the negative control. Genotyping can be done anytime after the organs have been put into culture. Extracted DNA can be used immediately or stored at −20°C.

10. Vortex the tubes frequently. Make sure that the cap is tightly closed and be extremely cautious with the hot liquid when opening the tubes. The tissue is considered fully digested when the tissue fragments are completely dissolved in the solution.

11. We do not need to genotype for *YFP* allele in the progenies from *Gli1^{Cre/+}* × *Rosa26^{YFP/YFP}*, since the genotype of all the progenies is *Rosa26^{YFP/+}* (see Note 9).

12. Immunohistochemistry is performed on the ovary complexes collected at 18.5 dpc (time zero) and the ovary complexes after a 72-h culture. At 18.5 dpc, *YFP*-positive cells are mostly present in the mesonephros. After 72 h of culture, *YFP*-positive cells appear in the ovary, indicating that these *YFP*-positive cells came from the mesonephros. By double staining the section with granulosa cell marker FOXL2, we can identify whether these mesonephros-derived *Gli1*-expressing cells contribute to FOXL2-positive granulosa cells or FOXL2-negative mesenchymal cells in ovary. If *Rosa26LacZ* reporter is used instead of *Rosa26YFP*, then *LacZ* primary antibody should be included along with the FOXL2 antibody to mark the *LacZ*-positive cells.

Acknowledgments

This research was supported by the National Institute of Health (HD046861, HD059661, and ES018163). It was also supported in part by the Intramural Research Program of the National Institute of Environmental Health Sciences (NIEHS) and NIH Graduate Partnership Program.

References

1. Stern CD, Fraser SE (2001) Tracing the lineage of tracing cell lineages. Nat Cell Biol 3: E216–218.

2. Gu G, Brown JR, Melton DA (2003) Direct lineage tracing reveals the ontogeny of pancreatic cell fates during mouse embryogenesis. Mech Dev 120:35–43.

3. Feil R, Brocard J, Mascrez B, LeMeur M, Metzger D, Chambon P (1996) Ligand-activated site-specific recombination in mice. Proc Natl Acad Sci USA 93:10887–10890.

4. Kellendonk C, Tronche F, Monaghan AP, Angrand PO, Stewart F, Schutz G (1996)

Regulation of Cre recombinase activity by the synthetic steroid RU 486. Nucleic Acids Res 24:1404–1411.

5. Song DL, Chalepakis G, Gruss P, Joyner AL (1996) Two Pax-binding sites are required for early embryonic brain expression of an Engrailed-2 transgene. Development 122:627–635.

6. Zinyk DL, Mercer EH, Harris E, Anderson DJ, Joyner AL (1998) Fate mapping of the mouse midbrain-hindbrain constriction using a site-specific recombination system. Curr Biol 8:665–668.

7. Bai CB, Auerbach W, Lee JS, Stephen D, Joyner AL (2002) Gli2, but not Gli1, is required for initial Shh signaling and ectopic activation of the Shh pathway. Development 129:4753–4761.

8. Ahn S, Joyner AL (2004) Dynamic changes in the response of cells to positive hedgehog signaling during mouse limb patterning. Cell 118:505–516.

9. Huang CC, Miyagawa S, Matsumaru D, Parker KL, Yao HH Progenitor cell expansion and organ size of mouse adrenal is regulated by sonic hedgehog. Endocrinology 151:1119–1128.

10. Zawadzka M, Rivers L E, Fancy SP, Zhao C, Tripathi R, Jamen F, Young K, Goncharevich A, Pohl H, Rizzi M, Rowitch DH, Kessaris N, Suter U, Richardson WD, Franklin RJ. CNS-resident glial progenitor/stem cells produce Schwann cells as well as oligodendrocytes during repair of CNS demyelination. Cell Stem Cell 6:578–590.

11. Wijgerde M, Ooms M, Hoogerbrugge JW, Grootegoed JA (2005) Hedgehog signaling in mouse ovary: Indian hedgehog and desert hedgehog from granulosa cells induce target gene expression in developing theca cells. Endocrinology 146:3558–3566.

12. Russell MC, Cowan RG, Harman RM, Walker AL, Quirk SM (2007) The hedgehog signaling pathway in the mouse ovary. Biol Reprod 77:226–236.

13. Ren Y, Cowan RG, Harman RM, Quirk SM (2009) Dominant activation of the hedgehog signaling pathway in the ovary alters theca development and prevents ovulation, Mol Endocrinology 23:711–723.

14. Soriano P (1999) Generalized lacZ expression with the ROSA26 Cre reporter strain. Nat Genet 21:70–71.

15. Srinivas S, Watanabe T, Lin CS, William CM, Tanabe Y, Jessell TM, Costantini F (2001) Cre reporter strains produced by targeted insertion of EYFP and ECFP into the ROSA26 locus. BMC Dev Biol 1:4.

16. Indra AK, Warot X, Brocard J, Bornert JM, Xiao JH, Chambon P, Metzger D (1999) Temporally-controlled site-specific mutagenesis in the basal layer of the epidermis: comparison of the recombinase activity of the tamoxifen-inducible Cre-ER(T) and Cre-ER(T2) recombinases. Nucleic Acids Res 27:4324–4327.

17. Friedrich G, and Soriano P (1991) Promoter traps in embryonic stem cells: a genetic screen to identify and mutate developmental genes in mice. Genes Dev 5:1513–1523.

18. Morris RJ, Liu Y, Marles L, Yang Z, Trempus C, Li S, Lin JS, Sawicki JA, Cotsarelis G (2004) Capturing and profiling adult hair follicle stem cells. Nat Biotechnol 22:411–417.

19. Danielian PS, Muccino D, Rowitch DH, Michael SK, McMahon AP (1998) Modification of gene activity in mouse embryos in utero by a tamoxifen-inducible form of Cre recombinase. Curr Biol 8:1323–1326.

20. Robinson SP, Langan-Fahey SM, Johnson DA, Jordan VC (1991) Metabolites, pharmacodynamics, and pharmacokinetics of tamoxifen in rats and mice compared to the breast cancer patient. Drug Metab Dispos 19:36–43.

21. Zervas M, Millet S, Ahn S, Joyner AL (2004) Cell behaviors and genetic lineages of the mesencephalon and rhombomere 1. Neuron 43:345–357.

Chapter 17

DNA Methylation Analysis of Germ Cells by Using Bisulfite-Based Sequencing Methods

Hisato Kobayashi and Tomohiro Kono

Abstract

Dynamic changes in DNA methylation at the gene-specific and genome-wide level occur during mammalian germ-cell development. However, the details of how the methylation profiles change remain largely unknown. Bisulfite sequencing analysis is a powerful technique to determine the methylation status of DNA at individual cytosine-guanine dinucleotide (CpG) sites and requires only a small amount of DNA for analysis. Here, we introduce two methods for bisulfite-based DNA methylation analyses using small samples such as germ cells: bisulfite Sanger sequencing at a specific locus and high-throughput bisulfite sequencing at the whole genome level.

Key words: DNA methylation, Mouse oocyte, Bisulfite sequencing, Bisulfite shotgun sequencing, Epigenome

1. Introduction

DNA methylation in vertebrates involves the addition of a methyl group to the carbon-5 position of cytosine residues, which occurs almost exclusively in the context of CpG (cytosine followed by guanine) dinucleotides. The cytosine methylation is a major epigenetic modification that plays vital roles in the activation, silencing, and maintenance of gene expression. In mammals, CpG islands (CGIs) are prominent owing to their high-density of CpG dinucleotides and frequent association with genes (often mapping to promoters or first exons). Recently, intensive research into CpG islands has expanded our understanding of the functional aspects of DNA methylation in higher eukaryotic processes such as cellular differentiation, regulation of genomic imprinting, X chromosome inactivation, tumorigenesis, and silencing of retrotransposons (1).

Wai-Yee Chan and Le Ann Blomberg (eds.), *Germline Development: Methods and Protocols*, Methods in Molecular Biology, vol. 825, DOI 10.1007/978-1-61779-436-0_17, © Springer Science+Business Media, LLC 2012

Throughout germline development, dynamic DNA methylation changes occur in a sex- and sequence-specific manner and, result in the establishment of unique methylation patterns in repetitive elements and sex-specific methylation imprints. Hence, the proper regulation of CpG dinucleotide methylation is indispensable for functional gamete development (2–6).

The nature of the epigenetic modifications responsible for the regulation of cell differentiation and embryo development has been extensively detailed using several approaches. One approach uses cleavage with methylation-sensitive restriction enzymes followed by two-dimensional gel electrophoresis (7) or high-density oligonucleotide arrays (8). A second approach is based on the enrichment of methylated DNA fragments by chromatin immunoprecipitation (ChIP) with the anti-methyl-CpG-binding domain protein (MBD-ChIP) or anti-methyl-cytosine antibodies (methylated DNA immunoprecipitation, MeDIP), followed by hybridization to arrays (9, 10). Recently, the enzyme digestion and ChIP approaches have been combined with high-throughput sequencing technologies, namely, Methyl-seq, MBD-seq, and MeDIP-seq analysis (11–13). These techniques have advantages over the use of arrays, for example, they yield sequence-level information that aids in distinguishing highly similar sequences. However, these methodologies can be limited by sequence-specific bias; enrichment of sites containing relatively high levels of cytosine methylation; reduction in complexity through only a small fraction of the genome being analyzed; and requiring at least microgram quantities of DNA, which is not feasible for samples with limited DNA amounts (e.g., clinical samples, specific tissues regions, and mammalian oocytes).

Bisulfite conversion is the third method utilized to investigate DNA methylation profiles. It can detect methylated or unmethylated cytosines at the resolution of a single CpG (14). Sodium bisulfite is used to convert unmethylated cytosine residues to uracil residues in single-stranded DNA. Three sequential reactions result in this conversion: sulfonation of cytosine to cytosine-6-sulfonate, deamination to uracil-6-sulfonate, and desulfonation to uracil. However, since 5-methylcytosine residues are nonreactive, they remain unconverted. Hence, the methylated and unmethylated cytosines can be distinguished upon sequencing, and the methylation pattern profiled with precision.

The researcher often has to choose between bisulfite sequencing for high resolution using small quantities of material or low resolution forglobal analysis using large sample sizes. Recently, shotgun bisulfite sequencing (BS-seq) with the Illumina Genome Analyzer has been used to determine whole genome CpG methylation sites; however, a large DNA sample is necessary for the analysis (15–18). Because of these technical limitations, a big issue in the study of the epigenome is that a standard platform for evaluating the methylome has not yet been provided. Obviously, the best standard is a high resolution, genome-wide profiling of the methylome

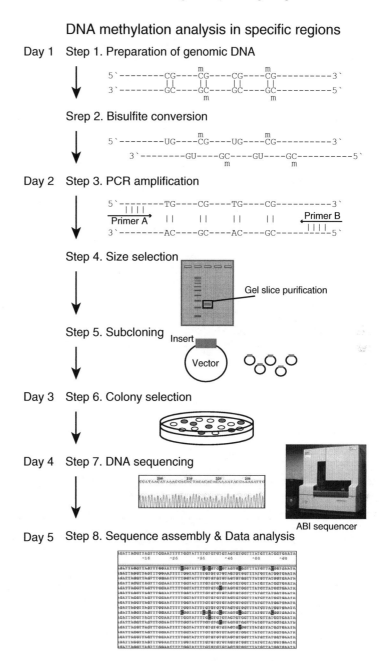

Fig. 1. Flow chart of bisulfite Sanger sequencing protocol to determine methylation patterns at specific sites in a few cells.

in oocytes and sperm. Notably, shotgun BS-seq enables the determination of the methylation status of individual CpG sites at the whole genome level without a bias toward CpG-rich regions. Moreover, the procedure embraces the possibility that the results can be achieved using relatively small samples. Here, we explain the original bisulfite Sanger sequencing method for the analysis of CpG-rich and non-CpG-rich loci in genomic DNA (Fig. 1) and the

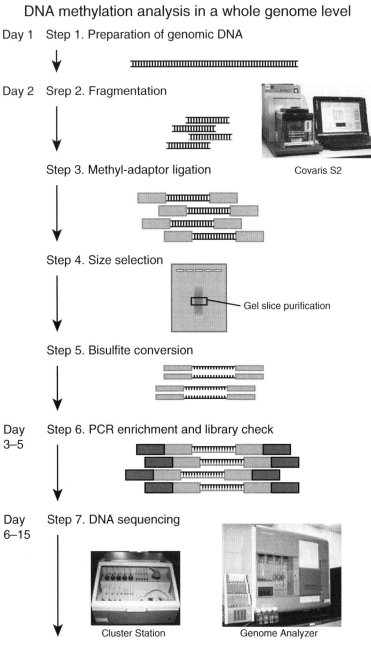

Fig. 2. Flow chart of shotgun bisulfite sequencing protocol to determine genome-wide methylation patterns.

shotgun BS-seq for whole genome level analysis (Fig. 2) of mouse germ cells. In addition, these methods potentially enable measurements of methylation profiles with other small samples, such as a small section of a clinical sample.

2. Materials

2.1. Sample

1. Germinal vesicle (GV) stage oocytes are prepared as described in ref 19. At least 50 oocytes are used in bisulfite Sanger sequencing and ≥5,000 oocytes for shotgun BS-seq.

2. For diploid cells, tissues, or testicular cells, prepare twice as many cells, i.e., ≥100 cells for bisulfite Sanger sequencing and ≥10,000 cells for shotgun BS-seq.

2.2. Bisulfite Sequencing by the Sanger Dideoxy Method

1. Lysis solution: 1 mM sodium dodecyl sulfate (SDS) and 280 μg/mL Proteinase K in deionized distilled water. Store at room temperature (RT).

2. *Escherichia coli* tRNA (Roche Diagnostics).

3. Epitect Bisulfite Kit (Qiagen).

4. TaKaRa Taq HotStart Version containing TaKaRa Taq HS, 10× polymerase chain reaction (PCR) buffer, and 2.5 mM dNTP mixture (see Note 1).

5. External and internal forward/reverse primer sets for two rounds of bisulfite-PCR (bisulfite nested PCR; see Note 2).

6. 2% (w/v) Agarose gel prepared by dissolving 2 g Agarose S in 100 mL 1× Tris/acetate/EDTA (TAE) buffer, 40 mM Tris–HCl; 20 mM acetate; 2 mM EDTA; pH 8.0.

7. 100 bp DNA ladder.

8. 10× Loading buffer.

9. 1× TAE buffer with 0.5 mg/mL ethidium bromide for DNA staining.

10. Wizard SV gel and PCR Clean-Up System.

11. pGEM-T Easy Vector System I containing pGEM-T Vector (50 ng/μL), T4 DNA Ligase, and 2× Rapid Ligation buffer.

12. *Escherichia coli* JM109 or DH5α competent cells.

13. LB agar plates: 2% (w/v) Bacto-tryptone, 1% (w/v) Yeast extract, 2% (w/v) NaCl, 1% (w/v) Agar in distilled water, autoclave at 121°C for 20 min. After autoclaving, cool to 60–80°C and add 100 μg/mL ampicillin. Pour immediately into 90 mm plates and store at 4°C until required (up to 1 month).

14. 20 mg/mL 5-Bromo-4-chloro-3-indolyl-β-d-galactopyranoside (X-gal) solution: Dissolve 100 mg X-gal in 5 mL of *N*, *N*-dimethylformamide in a 15-mL sterile polypropylene tube. Wrap the tube in aluminum foil to prevent damage by light and store at −20°C.

15. 200 mg/mL Isopropyl-β-d-thiogalactoside (IPTG) solution: Dissolve 2 g of IPTG in 8 mL of deionized distilled water in a sterile 15-mL tube. Adjust the volume to 10 mL with distilled water and filter through a 0.22 μm syringe filter into 1 mL aliquots and store at –20°C.

16. LB medium: 2% (w/v) Bacto-tryptone, 1% (w/v) yeast extract, 2% (w/v) NaCl, in distilled water, autoclave at 121°C for 20 min. After autoclaving, add 100 μg/mL ampicillin and store at 4°C until required (up to 1 month).

17. GenElute Plasmid Miniprep Kit.

18. Bigdye Terminator v3.1 Cycle Sequencing Kit.

19. Standard primers (e.g., SP6, 5′-GATTTAGGTGACACTAT AG-3′ and/or T7, 5′-TAATACGACTCACTATAGGG-3′).

20. 3 M Sodium acetate (pH 5.2).

21. 70% and 100% Ethanol.

22. Hi-Di formamide (Applied Biosystems).

23. ABI sequencer ABI 3730xl.

2.3. Bisulfite Shotgun Sequencing

1. QIAamp DNA Micro Kit (Qiagen).

2. Covaris S2 system (Covaris) for DNA fragmentation.

3. ChIP-Seq DNA Sample Prep Kit (Illumina).

4. Methylation Adaptor Oligo Mix (Illumina).

5. 3% (w/v) Agarose gel: prepare by dissolving 3 g Agarose S in 100 mL 1× TAE buffer.

6. 1× TAE buffer with 0.5 mg/mL ethidium bromide.

7. 50 bp DNA Ladder.

8. Epitect Bisulfite Kit.

9. QIAquick Gel Extraction Kit (Qiagen).

10. TaKaRa Taq HotStart Version.

11. Paired End Oligo Only Kit (Illumina) containing PE Primer 1.0 and 2.0 for Illumina sequencing.

12. Illumina Genome Analyzer II.

13. Standard and Paired-End Cluster Generation Kits v4 (Illumina).

14. 36-Cycle Sequencing Kit v4 (Illumina).

3. Methods

3.1. Bisulfite Sanger Sequencing

Bisulfite Sanger sequencing is widely used for DNA methylation analysis. It typically involves an initial treatment of genomic DNA with sodium bisulfite and a subsequent PCR amplification of specific regions of interest (e.g., CGIs and/or promoter regions). An overview of the methodology is provided in Fig. 1. It is recognized that this method is impractical for genome-wide DNA methylation analysis.

1. Resuspend ≥50 GV oocytes in 18 μL of lysis solution with 2 μg *E. coli* tRNA. Incubate the samples for 30–90 min at 37°C and then 15 min at 98°C (see Note 3).

2. Treat the lysate with sodium bisulfite using the Epitect Bisulfite Kit according to the manufacturer's instructions (protocol for low concentration of DNA and by centrifugation) (see Note 4). Elute bisulfite-converted DNA with 20 μL of EB buffer provided in the Epitect Bisulfite Kit, and 10 μL of the DNA solution is used for subsequent PCR step.

3. For the first round of PCR, the reaction conditions are as follows: a 20-μL volume reaction mixture containing 1.0 U of TaKaRa Taq HS, 1× PCR buffer, 25 μM dNTP mixture, 0.5 μM external forward primer, 0.5 μM external reverse primer. The PCR amplification is carried out with denaturation at 94°C for 1 min, followed by 15–25 cycles of amplification at 94°C for 30 s, 55–60°C for 30 s, 72°C for 30 s, and a final extension for 5 min at 72°C. Purification of PCR product is not required.

4. For the second round of PCR, use 2 μL of the first-reaction product. The reaction conditions for the second round of amplification are as follows: a 20-μL volume reaction mixture containing 1.0 U of TaKaRa Taq HS, 1× PCR buffer, 25 μM dNTP mixture, 0.5 μM internal forward primer, 0.5 μM internal reverse primer. The PCR is carried out with a denaturation at 94°C for 1 min, 25–30 cycles of amplification at 94°C for 30 s, 55–60°C for 30 s, 72°C for 30 s, and a final extension for 5 min at 72°C.

5. To purify the PCR products, separate the DNA fragments on a 2% agarose gel stained with ethidium bromide by electrophoresis. When the PCR is finished, add 2 μL of 10× Loading buffer to 20 μL PCR product. For sizing of the DNA, use 1 μL of 100 bp DNA ladder. Load each sample onto a 2% agarose gel. Run the gel at a constant 100 V for 30–40 min. Stain the gel in 1× TAE buffer with 0.5 mg/mL ethidium bromide for 15–30 min after electrophoresis (see Note 5). Cut out the region of the agarose gel containing the PCR amplified product

with a scalpel blade (~300 mg of gel slice) on a UV illuminator and purify the DNA with the Wizard SV gel and PCR Clean-Up System according to the manufacturer's instructions. Elute PCR products in 30 µL of nuclease-free water provided in the Wizard SV gel and PCR Clean-Up System.

6. Subclone all of the amplified, purified DNA into the pGEM-T vector according to the manufacturer's instruction. After subcloning, transform *E. coli* JM109 or DH5α competent cells by adding 2 µL of the DNA:vector product and placing the mixture on ice for 15 min. Incubate at 42°C for 45 s followed by ice for 2 min. Spread 30 µL of X-gal solution and 20 µL of IPTG solution onto LB agar plates containing ampicillin. Plate 50–100 µL of transformed cells onto the LB agar plates containing ampicillin/X-gal/IPTG, and incubate the plates at 37°C overnight.

7. Transfer >10 positive (white) colonies to 2 mL LB medium containing ampicillin in capped 15-mL tubes using sterile toothpicks (or a pipette tip) and shake at 37°C overnight at 150 rpm. Isolate plasmid DNA using a GenElute Plasmid Miniprep Kit according to the manufacturer's instructions (see Note 6).

8. Perform sequencing reactions using a BigDye terminator cycle sequencing kit with standard primers (e.g., 0.16 µM SP6 and/ or T7) according to the standard manufacturer's protocol.

9. Transfer each sequencing reactions to a clean 1.5 mL microcentrifuge tube containing the following: 2 µL of 3 M sodium acetate and 50 µL of 100% ethanol vortex briefly, and microcentrifuge at maximum speed for 5 min. Carefully aspirate the supernatant by pipetting and discard. Rinse the pellet with 100 µL of 70% ethanol. Microcentrifuge at maximum speed (>10,000×g) for 5 min, and carefully aspirate the supernatant again. Air-dry pellet for approximately 30 min at RT and dissolve the dried pellet in 15 µL of Hi-Di formamide. Analyze cDNA with an ABI sequencer.

10. Sequence reads from the cDNA of each clone are groomed manually and assembled (see Note 7).

3.2. High-Throughput Bisulfite Sequencing

The Genome Analyzer is a next-generation sequencing platform from Illumina for analyzing nucleic acid libraries at the gigabase level. The high-throughput sequencing technology is capable of generating approximately a 100 million raw (unidirectional) or paired-end (bidirectional) reads of 36–150 nt length per run, of which 50–70% typically pass through quality filtering. The massive parallel sequencing of bisulfite-converted genomic DNA enables the DNA methylation status to be determined at the whole genome level.

1. Extract genomic DNA from ≥5,000 GV oocytes using the QIAamp DNA Micro Kit (see Note 8). Elute the DNA in 80–100 μL of the provided Elution buffer.

2. Shear the 80–100 μL of each DNA sample into 100-bp fragments using the Covaris S2 system according to the manufacturer's instruction (see Note 9).

3. Ligate cytosine-methylated adapters for Genome Analyzer sequencing to the sheared DNA fragments using a ChIP-Seq DNA Sample Prep Kit according to the manufacturer's instruction (see Note 10). In the adapter ligation step, use Methylation Adaptor Oligo Mix instead of the (unmethylated) Adaptor Oligo Mix provided in the ChIP-Seq Kit. Elute adapter-ligated DNA fragments in 20 μL of EB buffer.

4. Add 2 μL of 10× Loading buffer to the adapter-ligated fragments; 1 μL of 50 bp DNA ladder is used for sizing. Load each sample onto a 3% agarose gel. Run the gel at a constant 100 V for 35–45 min. Stain the gel in 1× TAE buffer with 0.5 mg/mL Ethidium bromide for 15–30 min after electrophoresis (see Note 5). Cut out ~200-bp fragments (see Note 11) from the agarose gel with a scalpel blade on a UV illuminator and purify them using the QIAquick Gel Extraction Kit (see Note 11). Elute the DNA fragments in 40 μL of EB buffer.

5. Treat the isolated DNA fragments with sodium bisulfite using the Epitect Bisulfite Kit according to the manufacturer's instructions (protocol for low concentration of DNA and by centrifugation) (see Note 8) and elute the bisulfite-converted DNA in 20 μL of EB buffer, and 10 μL of the DNA solution is used for the subsequent PCR step.

6. Amplify the bisulfite-converted, adapter-ligated DNA by PCR using the following reaction mixture: 2.5 U of TaKaRa Taq Hot Start DNA polymerase, 5 μL 10× PCR buffer, 25 μM dNTPs, 1 μL PE Primer 1.0, 1 μL PE Primer 2.0 (50 μL final). The thermocycling parameters are as follows: 94°C for 1 min, then 20- to 25-cycle at 94°C for 30 s, 65°C for 30 s, and 72°C for 30 s, and then a final extension at 72°C for 5 min (see Note 12). Purify the reaction products using the MinElute Gel Extraction Kit and elute them in 20 μL of the EB buffer provided in the MinElute Kit (see Note 13). Use the Agilent 2100 Bioanalyzer for quality control of the libraries. Load 1 μL of resuspended construct on the Bioanalyzer, and check the size, purity, and concentration of the sample (see Note 14).

7. Sequence the library on a GAII (Illumina). Perform sample preparation, cluster generation, and sequencing by using the GAII Standard and Paired-End Cluster Generation Kits (Illumina) and the 36-Cycle Sequencing Kits according to the manufacturer's protocols (20) (see Note 15).

8. Process the sequenced reads using the Illumina standard base-calling pipeline. Map the generated sequence tags onto the mouse genome by following the procedure described in previous studies (15, 17), using the original, customized Perl program. Briefly, temporarily replace the cytosines in all tags with thymines. Then, align these tags to two in silico-converted mouse genome reference sequences. In the first reference sequence, the cytosines in the first strand ("Fwd" strand") have been converted to thymines. In the second reference sequence, the cytosines in the second strand ("Rev" strand") have been converted to adenines. Finally, compile all aligned tags (those tags that match up at 32-nt perfectly) to both "Fwd" and "Rev" strands (see Note 16).

4. Notes

1. We recommend using hot start Taq DNA polymerase to prevent primer dimer formation.

2. For amplification of small amount of bisulfite-treated DNA, we recommend "nested PCR," in two separate rounds of amplifications, using a pair of "external" primers (first round) and a pair of "internal" primers (second round) to increase the specificity of DNA amplification and reduce background due to nonspecific amplification. Prior to designing each primer, use the WORD program to convert the Cs to Ts in the target genomic DNA sequences except for those in CG dinucleotides. Design primers without CG dinucleotides. Each primer length should be 20–30 nt, and the amplified region should be no more than 300 bp (at least by internal primers). The best annealing temperature is usually, but not always, at 55–60°C. In addition, we recommend the MethPrimer Web site (http://www.urogene.org/methprimer/index1.html) for designing primers for methylation PCR. If possible, you should check whether your primers can be used for bisulfite sequencing by testing them on other enough amounts of DNA samples.

3. If you want highly purified DNA, we recommend using the QIAamp DNA Micro Kit (Qiagen) for DNA isolation, but it is requires the use of at least 200 GV oocytes, according to our experience.

4. We confirmed that these kits could work with our protocol: CpGenome First DNA Modification Kit (Millipore, Bedford, MA, Cat. No. S7824) and EZ DNA Methylation Kit (Zymo Research, Orange, CA, Cat. No. D5001). However, we do not recommend a specific kit for bisulfite conversion.

5. Ethidium bromide is a hazardous chemical that requires special storage, handling, and disposal as hazardous waste.

6. You can bypass culturing and subsequent purification steps using illustra TempliPhi DNA Amplification Kit (GE Healthcare). TempliPhi-generated circular DNA can be used directly for DNA sequencing.

7. The assembling of sequence reads can be performed using ATGC (Genetyx, Tokyo, Japan), Sequencher (Gene Codes, Ann Arbor, MI), or BioEdit software (http://www.mbio.ncsu.edu/BioEdit/BioEdit.html). In addition, BiQ analyzer and QUMA support alignment and visualization of DNA methylation data from bisulfite sequencing. Examples for the visualization of the bisulfite sequencing results were shown in our papers (21, 22). Generally, methylated and unmethylated CpG sites were designed as filled and empty circles.

8. We recommend the use of carrier RNA. Then, RNase treatment is not required for library construction.

9. The Nebulizer (Invitrogen) and dsDNA Fragmentase (New England Biolabs) are also effective for generating fragments, but DNA purification by QIAquick column is required. We consider that Covaris shearing works the best with nanogram quantities of DNA.

10. At the adapter ligation step, use cytosine-methylated adapters instead of the unmethylated-adapter oligo mix attached to the Sample Prep Kit.

11. Ethidium bromide staining cannot visualize sheared genomic DNA under the order of a microgram. We recommend using a 50-bp ladder to estimate the proper position for excising a gel band.

12. It is better to use PCR-enriched fragments amplified from a small number of PCR cycles.

13. We recommend checking whether the PCR libraries are available for Illumina sequencing, i.e., completely bisulfite-converted by subcloning (same as steps 6–9 of Subheading 3.1). At least 50 colonies are required from each library for sequencing.

14. The purified product should be a distinct band at approximately 200 bp. The minimum molarity is 10 nmol/L. If there are nonspecific amplifications or primer dimers, separate the DNA fragments by gel electrophoresis and purify them using the QIAquick Gel Extraction Kit.

15. The procedure is often updated with the kit version. You should carry out these steps according to the latest manufacturer's instructions.

16. Currently, Illumina does not support bisulfite sequence alignment.

References

1. Jaenisch R, Bird A (2003) Epigenetic regulation of gene expression: how the genome integrates intrinsic and environmental signals. Nat Genet 33 Suppl:245–254.

2. Kono T, Obata Y, Wu Q, Niwa K, Ono Y, Yamamoto Y, Park ES, Seo JS, Ogawa H (2004) Birth of parthenogenetic mice that can develop to adulthood. Nature 428: 860–864.

3. Bourc'his D, Xu GL, Lin CS, Bollman B, Bestor TH (2001) Dnmt3L and the establishment of maternal genomic imprints. Science 294: 2536–2539.

4. Hata K, Okano M, Lei H, Li E (2002) Dnmt3L cooperates with the Dnmt3 family of de novo DNA methyltransferases to establish maternal imprints in mice. Development 129:1983–1993.

5. Kaneda M, Okano M, Hata K, Sado T, Tsujimoto N, Li E, Sasaki H (2004) Essential role for de novo DNA methyltransferase Dnmt3a in paternal and maternal imprinting. Nature 429:900–903.

6. Hirasawa R, Chiba H, Kaneda M, Tajima S, Li E, Jaenisch R, Sasaki H (2008) Maternal and zygotic Dnmt1 are necessary and sufficient for the maintenance of DNA methylation imprints during preimplantation development. Genes Dev 22: 1607–1616.

7. Costello JF, Fruhwald MC, Smiraglia DJ, Rush LJ, Robertson GP, Gao X, Wright FA, Feramisco JD, Peltomaki P, Lang JC, Schuller DE, Yu L, Bloomfield CD, Caligiuri MA, Yates A, Nishikawa R, Su Huang H, Petrelli NJ, Zhang X, O'Dorisio MS, Held WA, Cavenee WK, Plass C (2000) Aberrant CpG-island methylation has non-random and tumour-type-specific patterns. Nat Genet 24:132–138.

8. Lippman Z, Gendrel AV, Colot V, Martienssen R (2005) Profiling DNA methylation patterns using genomic tiling microarrays. Nat Methods 2:219–224.

9. Weber M, Davies JJ, Wittig D, Oakeley EJ, Haase M, Lam WL, Schubeler D (2005) Chromosome-wide and promoter-specific analyses identify sites of differential DNA methylation in normal and transformed human cells. Nat Genet 37: 853–862.

10. Zhang X, Yazaki J, Sundaresan A, Cokus S, Chan SW, Chen H, Henderson IR, Shinn P, Pellegrini M, Jacobsen SE, Ecker JR (2006) Genome-wide high-resolution mapping and functional analysis of DNA methylation in arabidopsis. Cell 126:1189–1201.

11. Down TA, Rakyan VK, Turner DJ, Flicek P, Li H, Kulesha E, Graf S, Johnson N, Herrero J, Tomazou EM, Thorne NP, Backdahl L, Herberth M, Howe KL, Jackson DK, Miretti MM, Marioni JC, Birney E, Hubbard TJ, Durbin R, Tavare S, Beck S (2008) A Bayesian deconvolution strategy for immunoprecipitation-based DNA methylome analysis. Nat Biotechnol 26:779–785.

12. Brunner AL, Johnson DS, Kim SW, Valouev A, Reddy TE, Neff NF, Anton E, Medina C, Nguyen L, Chiao E, Oyolu CB, Schroth GP, Absher DM, Baker JC, Myers RM (2009) Distinct DNA methylation patterns characterize differentiated human embryonic stem cells and developing human fetal liver. Genome Res 19:1044–1056.

13. Serre D, Lee B H, Ting AH (2009) MBD-isolated Genome Sequencing provides a high-throughput and comprehensive survey of DNA methylation in the human genome. Nucleic Acids Res 38:391–399.

14. Frommer M, McDonald LE, Millar DS, Collis CM, Watt F, Grigg GW, Molloy PL, Paul CL (1992) A genomic sequencing protocol that yields a positive display of 5-methylcytosine residues in individual DNA strands. Proc Natl Acad Sci USA 89:1827–1831.

15. Laurent L, Wong E, Li G, Huynh T, Tsirigos A, Ong CT, Low HM, Kin Sung KW, Rigoutsos I, Loring J, Wei CL. (2010) Dynamic changes in the human methylome during differentiation. Genome Res 20:320–331.

16. Lister R, O'Malley RC, Tonti-Filippini J, Gregory BD, Berry CC, Millar AH, Ecker JR (2008) Highly integrated single-base resolution maps of the epigenome in Arabidopsis. Cell 133:523–536.

17. Cokus SJ, Feng S, Zhang X, Chen Z, Merriman B, Haudenschild CD, Pradhan S, Nelson SF, Pellegrini M, Jacobsen SE (2008) Shotgun bisulphite sequencing of the Arabidopsis genome reveals DNA methylation patterning. Nature 452:215–219.

18. Lister R, Pelizzola M, Dowen RH, Hawkins RD, Hon G, Tonti-Filippini J, Nery JR, Lee L, Ye Z, Ngo QM, Edsall L, Antosiewicz-Bourget J, Stewart R, Ruotti V, Millar AH, Thomson JA, Ren B, Ecker JR (2009) Human DNA methylomes at base resolution show widespread epigenomic differences. Nature 462:315–322.

19. Kawahara M, Obata Y, Sotomaru Y, Shimozawa N, Bao S, Tsukadaira T, Fukuda A, Kono T (2008) Protocol for the production of viable bimaternal mouse embryos, Nat Protoc 3:197–209.

20. Bentley DR, Balasubramanian S, Swerdlow HP, Smith GP, Milton J, Brown CG, Hall KP, Evers DJ, Barnes CL, Bignell HR, Boutell JM, Bryant J, Carter RJ, Keira Cheetham R, Cox AJ, Ellis

DJ, Flatbush MR, Gormley NA, Humphray SJ, Irving LJ, Karbelashvili MS, Kirk SM, Li H, Liu X, Maisinger KS, Murray LJ, Obradovic B, Ost T, Parkinson ML, Pratt MR, Rasolonjatovo IM, Reed MT, Rigatti R, Rodighiero C, Ross MT, Sabot A, Sankar SV, Scally A, Schroth GP, Smith ME, Smith VP, Spiridou A, Torrance PE, Tzonev SS, Vermaas EH, Walter K, Wu X, Zhang L, Alam MD, Anastasi C, Aniebo IC, Bailey DM, Bancarz IR, Banerjee S, Barbour SG, Baybayan PA, Benoit VA, Benson KF, Bevis C, Black PJ, Boodhun A, Brennan JS, Bridgham JA, Brown RC, Brown AA, Buermann DH, Bundu AA, Burrows JC, Carter NP, Castillo N, Chiara E Catenazzi M, Chang S, Neil Cooley R, Crake NR, Dada OO, Diakoumakos KD, Dominguez-Fernandez B, Earnshaw DJ, Egbujor UC, Elmore DW, Etchin SS, Ewan MR, Fedurco M, Fraser LJ, Fuentes Fajardo KV, Scott Furey W, George D, Gietzen KJ, Goddard CP, Golda GS, Granieri PA, Green DE, Gustafson DL, Hansen NF, Harnish K, Haudenschild CD, Heyer NI, Hims MM, Ho JT, Horgan AM, Hoschler K, Hurwitz S, Ivanov DV, Johnson MQ, James T, Huw Jones TA, Kang GD, Kerelska TH, Kersey AD, Khrebtukova I, Kindwall AP, Kingsbury Z, Kokko-Gonzales PI, Kumar A, Laurent MA, Lawley CT, Lee SE, Lee X, Liao AK, Loch JA, Lok M, Luo S, Mammen RM, Martin JW, McCauley PG, McNitt P, Mehta P, Moon KW, Mullens JW, Newington T, Ning Z, Ling Ng B, Novo SM, O'Neill MJ, Osborne MA, Osnowski A, Ostadan O, Paraschos LL, Pickering L, Pike AC, Pike AC, Chris Pinkard D, Pliskin DP, Podhasky J, Quijano VJ, Raczy C, Rae VH, Rawlings SR, Chiva Rodriguez A, Roe PM, Rogers J, Rogert Bacigalupo MC, Romanov N, Romieu A, Roth RK, Rourke NJ, Ruediger ST, Rusman E, Sanches-Kuiper RM, Schenker MR, Seoane JM, Shaw RJ, Shiver MK, Short SW, Sizto NL, Sluis JP, Smith MA, Ernest Sohna Sohna J, Spence EJ, Stevens K, Sutton N, Szajkowski L, Tregidgo CL, Turcatti G, Vandevondele S, Verhovsky Y, Virk SM, Wakelin S, Walcott GC, Wang J, Worsley GJ, Yan J, Yau L, Zuerlein M, Rogers J, Mullikin JC, Hurles ME, McCooke NJ, West JS, Oaks FL, Lundberg PL, Klenerman D, Durbin R, Smith AJ (2008) Accurate whole human genome sequencing using reversible terminator chemistry. Nature 456:53–59.

21. Hiura H, Obata Y, Komiyama J, Shirai M, Kono T (2006) Oocyte growth-dependent progression of maternal imprinting in mice. Genes Cells 11:353–361.

22. Kobayashi H, Yamada K, Morita S, Hiura H, Fukuda A, Kagami M, Ogata T, Hata K, Sotomaru Y, Kono T (2009) Identification of the mouse paternally expressed imprinted gene Zdbf2 on chromosome 1 and its imprinted human homolog ZDBF2 on chromosome 2. Genomics 93:461–472.

Chapter 18

Gene Expression in Mouse Oocytes by RNA-Seq

Eric Antoniou and Robert Taft

Abstract

Second-generation sequencers such as the Illumina GAIIx make possible the study of all transcribed loci in a genome across an almost endless dynamic range. Although typical protocols call for starting from at least 1 μg of total RNA, this is not possible when studying small tissues or rare cell types. This chapter explains how to prepare Illumina sequencing libraries from mouse oocytes. The protocol is also suitable for mural and cumulus cells, flow sorted or laser captured cells.

Key words: Oocyte, RNA-Seq, Next-generation sequencing, Gene expression

1. Introduction

Since the advent of DNA microarray, it is possible to study gene expression of tens of thousands of genes simultaneously. However, DNA microarrays suffer from several limitations, namely, a limited dynamic range that compress the fold change of highly expressed genes, the inability to reliably measure the activity of genes expressed at low levels in the cells, and the need for a priori information on where the genes are in the genome of interest. This last limitation has become more striking through the last 6 years, as many studies have shown extensive transcription outside of known exons (1–4). The arrival of second-generation sequencing machines (Illumina GA and HiSeq, AB Solid, Roche 454) has opened the door to digital gene expression (RNA-Seq). Instead of relying on hybridization to predefined probes, it is now possible to sequence the cDNA molecules and directly count how many sequences come from a given gene. The number of sequences, after being normalized to account for experimental noise, is directly related to the level of expression of the genes. The dynamic range only depends

Wai-Yee Chan and Le Ann Blomberg (eds.), *Germline Development: Methods and Protocols*, Methods in Molecular Biology, vol. 825, DOI 10.1007/978-1-61779-436-0_18, © Springer Science+Business Media, LLC 2012

on how many total sequences are obtained. In other words, you can sequence a lot (200 millions sequences/sample) to study genes expressed at very low level in the cells or limit your sequencing to a minimum (eight to ten millions sequences/sample) if you are only interested in more commonly expressed genes. This chapter introduces a method to perform RNA-Seq on mouse oocytes. It is also applicable to other cell types in the ovarian follicle, such as cumulus or mural granulosa cells, or to any flow sorted or laser captured cells. The analysis of this type of data deserves another chapter by itself. Nevertheless, readers will find the following articles to be a good introduction in the analysis of RNA-Seq data (5–10).

2. Materials

2.1. For Oocytes Isolation

1. Modified Dulbecco's phosphate buffered saline (PBS) (see Note 1): dissolve 0.10 g $CaCl_2 \cdot 6H_2O$ in cell culture grade water. Dissolve 8.0 g NaCl, 0.20 g KCl, 1.15 Na_2HPO_4, 0.20 g KH_2PO_4, 0.10 g $MgCl_2 \cdot 6H_2O$, 1.0 g dextrose, 0.036 g sodium pyruvate, 0.075 g penicillin G, and 0.050 g streptomycin sulfate in 750 ml cell culture grade H_2O, then add the $CaCl_2 \cdot 6H_2O$ solution. Adjust pH to 7.4. Adjust volume to 1 L with additional cell culture grade H_2O. Filter sterilize and store at 4°C for up to 2 weeks. Add 3.0 g BSA, allowing the BSA to dissolve slowly without shaking (see Note 2).

2. Hyaluronidase: Create a 10 mg/ml stock solution by dissolving 50 mg of hyaluronidase in 5 ml modified DPBS. Filter sterilize, aliquot (30 μl/aliquot), freeze, and store at –20°C for months. To use, add 500 μl modified DPBS and mix gently.

3. Acid Tyrode's solution (see Note 3): dissolve 0.024 g $CaCl_2 \cdot 2H_2O_2$ in 5 ml H_2O.
   Dissolve 0.80 g NaCl, 0.20 g KCl, 0.0101 g $MgCl_2 \cdot 6H_2O$, and 0.40 g polyvinylpyrolidone (PVP) in 80 ml H_2O at room temperature. Add the dissolved $CaCl_2$ to the solution. Add water to bring the volume to about 95 ml and adjust pH to 2.5 with 1 N HCl. Bring to a final volume of 100 ml H_2O. Filter sterilize, aliquot, and store at –20°C.
   Solutions used but not described that should be added are:

4. PBS: 0.138 M NaCl, 0.0027 M KCl, 0.05% TWEEN® 20, pH 7.4, at 25°C.

5. 70% Ethanol: add 70 ml absolute ethanol to 30 ml molecular biology grade water (RNAse and DNAse free).

6. 2-Propanol (Isopropanol; $(CH_3)_2CHOH$ for RNA prep Sigma).

7. 80% Ethanol for RNA prep: add 80 ml absolute ethanol to 20 ml of molecular biology grade water (RNAse and DNAse free).

2.2. For Making the Sequencing Library

The following commercial products have been carefully validated with this protocol and are required for the successful completion of this protocol

1. Qiagen RNeasy micro extraction kit.

2. Qiagen QIAquick MinElute Purification kit.

3. Qiagen QIAquick PCR Purification kit.

4. NEBNext® DNA Sample Prep Reagent Set 1.

5. Ovation RNA-Seq for amplification of RNA (Nugen Inc.).

6. KAPA Library Quant Kits (http://www.kapabiosystems.com).

7. Agencourt RNAclean beads.

8. Absolute (100%) ethanol.

9. Beta-mercaptoethanol.

10. Ethidium bromide (3,8-diamino-5-ethyl-6-phenylphenanthri-dinium bromide) 10 mg/ml: add 1 ml of molecular biology grade water (RNAse and DNAse free) to 10 mg of ethidium bromide. Vortex for 30 s. Centrifuge for 30 s. Keep at 4°C.

11. Agarose gel 10× TRIS-HCl borate–EDTA (TBE) running buffer: 130 mM TRIS-HCl; 45 mM boric acid; 2.5 mM EDTA·Na$_2$ in H$_2$O, pH ~9.0 (25°C). Prior to agarose gel electrophoresis, dilute 100 ml of the 10× TBE with 900 ml of ultrapure water to obtain a working concentration of 1× TBE.

12. 2% Gel agarose gel: for a 100 ml gel, weigh 2.0 g of agarose in a 500 ml glass flask. Add 100 ml of 1× gel running buffer. Boil for 3 min in a microwave. Let the solution cool for 5 min and add 2 μl of ethidium bromide. Pour the solution into a gel tray.

13. Ladder for sizing: use 5 μl per well of the exACTGene* DNA Ladders from Fisher. It has DNA bands ranging from 25 to 650 bp in 50 bp increments.

14. 2-Propanol (Isopropanol).

3. Methods

3.1. Collection of Ovulated Oocytes and Preparation for RNA Extraction

3.1.1. Oocyte Collection

1. Beginning 13–14 h after the expected time of ovulation, euthanize (euthanasia must be carried out using a protocol approved by the Animal Care and Use Committee of your institution) two to five female mice of the same strain (see Note 4).

2. Place a female on its back on absorbent paper and generously spray the abdomen with 70% EtOH.

3. Expose the abdominal cavity by lifting the small fold of skin in the center of the abdomen with forceps, making a small cut

with sharp 5″ scissors then pulling the skin in the opposite direction (toward the head and tail) until the abdominal cavity is completely exposed.

4. Open the peritoneum by lifting a small fold of the peritoneum with microdissection forceps, making a small lateral cut using sharp–sharp 3″ scissors, then enlarging the cut to expose internal organs while being careful to avoid cutting any internal organs.

5. Push the intestines toward the head and reveal the horns of the uterus, the oviducts and ovaries.

6. Remove both oviducts. This is accomplished most easily by removing the ovary and the tip of the uterine horn along with the oviduct, carefully avoiding the fat pad. Place the tissue into a 35 × 10 mm culture dish containing room temperature PBS.

7. Repeat for each of the females.

8. Recover oocytes by using microdissecting forceps to immobilize an oviduct while using a beveled hypodermic needle to open the ampulla. Cumulus enclosed oocytes should be easily visible within the ampulla of the oviduct under low magnification (10×) when using a stereomicroscope with a transmitted light base.

9. Repeat until all clutches have been released from the ampulla.

3.1.2. Removal of Cumulus Cells and Zonae

1. Add 500 µl of room temperature hyaluronidase solution to a 35 × 10 mm dish.

2. Transfer the oocytes in a minimal volume to the dish containing the hyaluronidase using a 1 ml pipette.

3. Gently agitate by aspirating and dispensing the oocytes two to three times in the hyaluronidase solution.

4. When the oocytes begin to disperse, use a finely drawn glass pipette to transfer oocytes to a 35 × 10 mm dish containing 2.5 ml of PBS, being careful to transfer as few cumulus cells as possible (see Note 5).

5. Transfer the oocytes in a minimal volume to a second dish containing room temperature PBS, aspirate and dispense using a finely drawn glass pipette to help ensure complete removal and separation of cumulus cells.

6. Transfer small groups (20–50) of healthy oocytes in no more than 20 µl to a 400 µl drop of Acid Tyrode's at room temperature (see Note 6).

7. Gently mix by aspirating and dispensing using a 200 µl pipette and a wide bore pipette tip.

8. When zonae have dissolved, flood the dish with 4 ml of cold (4°C) PBS (see Note 7).

9. Wash oocytes by transferring in a minimal volume through three drops of PBS.

10. Repeat until all oocytes have been treated.

11. Transfer oocytes in a minimal volume into a 1.5 ml microcentrifuge tube, snap freeze, and store at a temperature of –70°C or less until use.

3.1.3. RNA Extraction
(Qiagen RNeasy Micro Kit)

1. Add 21 ml 100% ethanol to 9 ml nuclease-free water to make fresh 70% ethanol.

2. Add 24 ml 100% ethanol to 6 ml nuclease-free water to make fresh 80% ethanol.

3. Add 10 µl beta-mercaptoethanol to 1 ml of RLT buffer (you will need 350 µl/sample) (see Note 8).

4. Add 4 volumes of 100% ethanol to the RPE buffer.

5. To prepare the DNAse stock solution, add 550 µl of nuclease-free water to the DNAse powder.

6. Label two sets of 1.5 ml tubes and label one set of Qiagen columns (see Note 9).

7. Add 100 µl of the prepared RLT buffer into 1.5 ml tubes. Place the oocytes into the tube of RLT buffer. Vortex for 30 s. Quickly add an additional 250 µl of the RLT buffer to the lysate, for a total volume of 350 µl of RLT buffer.

8. Centrifuge the cell lysates at $11,000 \times g$ for 3 min then transfer the lysates to new 1.5 ml tubes by pipetting away the supernatant from the debris (supernatant only); bubbles are normal.

9. Add 1 volume (350 µl) of 70% ethanol to the lysates, mix well by pipetting up and down four times. Place an RNeasy spin column into a 2 ml collection tube and add the cell lysates, including any precipitate that may have formed, to the column.

10. Centrifuge the tubes for 15 s at $8,000 \times g$ with the caps closed. Discard the flow-through. Add 350 µl of RW1 buffer to the spin columns and centrifuge 15 s at $8,000 \times g$ with caps closed. Discard flow-through. Place the tubes on ice.

11. Add 10 µl of the DNase I stock solution (provided with the Qiagen kit) to 70 µl of the RDD buffer and mix by gently flicking of the tube. Spin down briefly on a tabletop centrifuge and put on ice. You will need 80 µl of this solution per sample.

12. Add 80 µl of DNase I + RDD solution directly onto the silica-gel membrane (see Note 10). Incubate for 15 min at room temperature.

13. Add 350 µl of RW1 buffer onto the spin column and centrifuge 15 s at $8,000 \times g$ with caps closed. Discard the flow-through and the collection tubes. Transfer the spin column to

a new 2.0 ml collection tube. Add 500 µl of RPE buffer to the columns. Centrifuge for 15 s at 8,000×*g* with the caps closed. Discard the flow-through.

14. Add 500 µl of 80% ethanol to the columns. Centrifuge for 2 min at 8,000×*g* with caps closed. Discard flow-through and collection tubes. Transfer the spin columns to new 2.0 ml collection tubes. Centrifuge at full speed in a microcentrifuge (≥10,000×*g*) with caps open for 5 min. Discard flow-through and collection tubes.

15. Transfer the columns to new 1.5 ml tubes and add 14 µl of RNase-free water directly to the silica-gel membrane. Close the tubes and centrifuge for 1 min at full speed in a microcentrifuge (≥10,000×*g*). Place the tubes on ice. (RNA samples can also be frozen at −80°C at this point).

3.2. RNA Amplification (NuGen) (See Note 11)

Flick to mix all enzymes, spin 2 s in a tabletop centrifuge and place on ice. Thaw all other reagents at room temperature, vortex for 2 s, spin for 2 s, and place on ice.

3.2.1. First-Strand Synthesis

Add 2 µl of A1 primer to a 0.2 ml PCR tube. Add 5 µl of total RNA sample (10 ng). Spin for 2 s and place on ice. Place the tube in a prewarmed thermal cycler for 5 min at 65°C, then cool to 4°C. Spin and place on ice. Add 2.5 µl of first-strand buffer and 0.5 µl of first-strand enzyme. Mix by pipetting six to eight times and spin for 2 s. Incubate in a thermo cycler using the following program:

4°C for 1 min
25°C for 10 min
42°C for 10 min
70°C for 15 min
Cool to 4°C. Spin for 2 s and place the tube on ice

3.2.2. Second-Strand cDNA Synthesis

Remove the Agencourt RNAclean purification beads from the fridge and let them equilibrate to room temperature for at least 15 min. Add 9.7 ml of second-strand buffer and 0.3 ml of second-strand enzyme to the cDNA tube. Place on ice. Mix by pipetting up and down six to eight times. Spin for 2 s. Incubate in a thermocycler using the following program:

4°C for 1 min
25°C for 10 min
50°C for 30 min
80°C for 20 min
Cool to 4°C. Spin for 2 s and place the tube on ice

3.2.3. Purification of Unamplified cDNA

Invert the container containing the Agencourt beads several times. Do not centrifuge. Add 32 ml of bead suspension to the tube containing the cDNA following second-strand synthesis; mix by pipetting ten times. Incubate at room temperature for 10 min (while waiting, make fresh 70% ethanol, 600 μl per reaction). Place the tube on the magnet and wait for 5 min to clear the solution of beads. Remove 42 μl of the buffer from the tube and discard. Add 200 μl of fresh 70% ethanol and incubate 30 s at room temperature. Remove the ethanol. Repeat the ethanol wash twice. Wait 1 min for the remaining ethanol to settle at the bottom of the tube. Remove the remaining ethanol carefully. Air dry the beads on the magnet for at least 15 min. Inspect the tubes to make sure the beads are dried.

3.2.4. SPIA Amplification

Thaw the SPIA enzyme mix on ice, mix by inverting five times, and spin for 2 s. Add 40 μl of the mix to the cDNA bound to dried beads. Set a pipette to 30 μl and mix well by pipetting ten times. You need to get most of the beads back in suspension and removed from the tube wall. Incubate in a thermo cycler using the following protocol:

4°C for 1 min
47°C for 1 h
95°C for 5 min
Cool to 4°C. Spin the tube for 2 s and place on ice. Move to the postamplification work area. Transfer the tube to a magnetic stand and wait for 5 min to clear the solution of beads

3.2.5. Post-SPIA Amplification Modifications

Transfer 35 μl of supernatant to a fresh tube, taking care to leave the beads behind. Add 5 μl of the primer mix E1. Mix by pipetting eight times and spin for 2 s. Place on ice. Incubate at 98°C for 3 min. Cool on ice. Add 5 μl of buffer mix E2 and 5 μl of the enzyme mix E3. Mix by pipetting eight times and spin for 2 s. Place on ice. Incubate in a thermo cycler using the following program:

4°C for 1 min
30°C for 10 min
42°C for 15 min
75°C for 10 min
Cool to 4°C
Spin the tube for 2 s and place on ice

3.3. Purification of Amplified Complementary DNA (acDNA) with the Qiagen Qiaquick PCR Purification Kit

Add 250 µl of PB buffer into a 1.5 ml tube. Pipette 50 µl of acDNA in the tube. Vortex for 5 s and spin down for 2 s. Load the 300 µl onto the QIAquick spin column. Centrifuge 1 min at >10,000×g. Discard the flow-through. Add 700 µl of fresh 80% ethanol. Centrifuge 1 min at >10,000×g. Discard the flow-through. Centrifuge the column for an additional minute at >10,000×g. Discard the flow-through. Tap the tip of the column onto a kim-wipe tissue in order to remove any residual 80% ethanol. Place the column into a new 1.5 ml tube. Add 30 µL of EB buffer to the center of each column. Incubate for 5 min at room temperature. Centrifuge for 1 min at >10,000×g. Collect the flow-through sample. Approximately 30 µL of acDNA should be available.

3.4. Making the Sequencing Library

This step uses Qiagen and NEB reagents

1. Dilute 500 ng (you can use as much as 2 µg and as little as 100 ng) of acDNA into 50 µl of RNase-free water, place in a 0.2 ml PCR tube. Add 10 µl of 10× end repair buffer, 5 µl of the end repair enzyme mix and 35 µl of nuclease-free water.

2. Flick the tube and centrifuge briefly. Incubate at 20°C for 30 min in a thermocycler.

3. Transfer the reaction mix to a 1.5 ml tube. Add 500 µl of buffer PB (QIAquick PCR Purification kit) to the reaction. Mix and centrifuge briefly. Transfer to a purification column and place the column in a collection tube. Spin at >10,000×g for 1 min. Discard the flow-through liquid. Invert the collection tube and tap gently on an absorbent paper towel to remove all liquid. Place the column into the same collection tube. Add 750 µl of PE buffer. Centrifuge for 1 min at >10,000×g. Discard the flow-through as described above. Replace the column in the same collection tube and centrifuge 1 min at >10,000×g. Place the column into a fresh 1.5 ml tube (cap open). Add 42 µl of buffer EB to the center of the column. Wait for 3 min and centrifuge 1 min at >10,000×g. Discard the column. Transfer the acDNA to a 0.2 ml PCR tube.

4. Add 5 µl of 10× A-tailing buffer to the eluted acDNA. Add 3 µl of Klenow fragment DNA polymerase (see Note 12), flick the tube and spin briefly. Incubate at 37°C for 30 min in a thermocycler.

5. Transfer the reaction mix to a 1.5 ml tube. Add 250 µl of buffer PB (QIAquick PCR Purification kit) to the reaction. Mix and centrifuge briefly. Transfer to a purification column and place the column in a collection tube. Spin at >10,000×g for 1 min. Discard the flow-through liquid. Invert the collection tube and tap gently on an absorbent paper towel to remove all liquid. Place the column into the same collection tube. Add 750 µl of PE buffer. Centrifuge for 1 min at >10,000×g.

Discard the flow-through as described above. Replace the column in the same collection tube and centrifuge 1 min at >10,000×g. Place the column into a fresh 1.5 ml tube (cap open). Add 38 μl of buffer EB to the center of the column. Wait for 3 min and centrifuge 1 min at >10,000×g. Discard the column. Transfer the eluted acDNA into a 0.2 ml PCR tube.

6. Add 10 μl of 10× quick reaction ligase buffer, 1 μl of quick T4 DNA ligase, and 1 μl of adapters (see Note 13). Incubate for 15 min at room temperature (22°C plus or minus 2°C).

7. Transfer the reaction to a new 1.5 ml tube. Add 250 μl of buffer PB (QIAquick MinElute Purification kit) to the reaction. Mix and centrifuge briefly. Transfer to a purification column (see Note 14) and place the column in a collection tube. Spin at >10,000×g for 1 min. Discard the flow-through liquid. Invert the collection tube and tap gently on an absorbent paper towel to remove all liquid. Place the column into the same collection tube. Add 750 μl of PE buffer. Centrifuge for 1 min at >10,000×g. Discard the flow-through as described above. Replace the column in the same collection tube and centrifuge 1 min at >10,000×g. Place the column into a fresh 1.5 ml tube (cap open). Add 10 μl of buffer EB to the center of the column. Wait for 3 min and centrifuge 1 min at >10,000×g. Discard the column. Keep the flow-through containing the adapter-ligated acDNA.

8. Prepare a 2% agarose gel using Seakem Agarose (Lonza). Add 5 μl of ethidium bromide (10 mg/μl) to the cooled gel and pour the gel. Add gel loading buffer to adapter-ligated acDNA. Load the gel and let at least two empty wells between each sample to avoid any cross-contamination. Load DNA ladder on both side of the gel. Run the gel at 100 V for 90 min.

9. Visualize the gel under a UV lamp. Make sure to use a long wavelength lamp (360 nm) as shorter wavelengths will damage the DNA and make it impossible to amplify by PCR (see Note 15). Use a GelCatcher™ (the gel company Inc.) to cut a thin gel slice containing 300 bp size DNA fragments. Place the GeneCatcher™ in a 1.5 ml tube and centrifuge briefly to recover the gel slice.

10. Weigh the gel slices (weight the tubes containing the gel slices, using an empty tube to set the balance at zero). Add 6 volumes of buffer QG for 1 volume of gel (QIAquick gel extraction kit). Incubate at room temperature until the gel slices are completely dissolved, vortexing for 5 s every 2 min. Add one gel volume of isopropanol to the mix (it increases the recovery rate of DNA fragments). Load the sample onto a column and centrifuge 1 min at >10,000×g. Discard the flow-through.

Invert the collection tube and tap gently on an absorbent paper towel to remove all liquid. Place the column into the same collection tube. Add 500 µl of QG buffer. Centrifuge at >10,000 × g for 1 min. Discard the flow-through. Invert the collection tube and tap gently on an absorbent paper towel to remove all liquid. Add 750 µl of PE buffer. Centrifuge for 1 min at >10,000 × g. Discard the flow-through as described above. Replace the column in the same collection tube and centrifuge 1 min at >10,000 × g. Place the column into a fresh 1.5 ml tube (cap open). Add 28 µl of buffer EB to the center of the column. Wait for 3 min and centrifuge 1 min at >10,000 × g. Discard the column. Transfer the adapter-ligated acDNA to a 0.2 ml PCR tube.

11. Add 30 µl of 2× Phusion HF buffer and 1 µl of each PCR primer (see Note 16). Place the tube in a thermocycler and use the following amplification program.

    (a) 95°C for 10 s

    (b) Then 15 cycles as follows:

        98°C for 10 s, 65°C for 30 s, 72°C for 30 s

    (c) 72°C for 5 min, then hold at 4°C

12. Transfer the reaction to a 1.5 ml tube. Add 250 µl of buffer PB (QIAquick PCR Purification kit) to the reaction. Mix and centrifuge briefly. Transfer to a purification column and place the column in a collection tube. Spin at >10,000 × g for 1 min. Discard the flow-through liquid. Invert the collection tube and tap gently on an absorbent paper towel to remove all liquid. Place the column into the same collection tube. Add 750 µl of PE buffer. Centrifuge for 1 min at >10,000 × g. Discard the flow-through as described above. Replace the column in the same collection tube and centrifuge 1 min at >10,000 × g. Place the column into a fresh 1.5 ml tube (cap open). Add 30 µl of buffer EB to the center of the column. Wait for 3 min and centrifuge 1 min at >10,000 × g. Discard the column.

3.5. Library Quality Control

Prior to sequencing the library, ensure the library does not contain too many primer dimers. Both the average size of the DNA fragments, the distribution (for paired-end sequencing) and an accurate measure of the quantity of cDNA fragments that have adapters at both ends need to be determined. Only cDNA with primers on both ends will be able to bind to the flow cell and be sequenced. We routinely use the following two procedures to achieve all this.

3.6. Agilent Bioanalyzer

The size, distribution, and primer dimer contamination are measured using a DNA 1000 chip on the Agilent Bioanalyzer. The chip is used according to the manufacturer's protocol. The peak at 120 bp is the primer dimer. Good libraries will have less than 10% primer dimer and a tight size distribution.

3.7. qPCR with Kapa Biosystems Kit

The Kapa Biosystem quantification qPCR is used to quantify the concentration of target fragments. The kit (KK4809) contains a modified DNA polymerase that can amplify sequences over a wide range of sizes and GC content. The kit is used according to Kapa Biosystem recommendations but we recommend diluting your libraries to 10^{-4}, 10^{-5} and 10^{-6}. Lower dilutions (10^{-3} or 10^{-2}) will often result in data points outside the range of the standard curve.

4. Notes

1. Modified Dulbecco's PBS is commercially available from several vendors in both powdered and liquid form.

2. It is not necessary to prepare 1 L batches; smaller batches may be more convenient. Likewise, it is not necessary to add BSA to the entire batch, but only to aliquots. Careful addition of BSA is essential as vigorous mixing will result in frothing and denaturing of the protein. This is also why the BSA is added after filtration, although BSA can be added prior to filtration if positive pressure filtration should be used to prevent foaming.

3. Acid Tyrode's solution is commercially available.

4. Often animals will be superovulated prior to oocyte collection. Superovulation increases yield and synchronizes ovulation. Typically, pregnant mare serum gonadotropin (2.5–5.0 IU by IP injection) is administered 62–58 h prior to the desired time of collection and human chorionic gonadotropin (2.5–5.0 IU by IP injection) is administered 12–14 h prior to desired time of collection. However, oocytes could be collected from natural cycles, although the ideal collection times are inconvenient as ovulation naturally occurs during the night. Alternatively, it may be desirable to collect oocytes prior to ovulation; these can be collected by collecting ovaries at the desired time point prior to ovulation Follicles are punctured on the ovaries with a needle to release oocytes and adherent cumulus cells; oocytes should be collected into calcium and magnesium free PBS to help loosen adherent cumulus cells. Once oocytes have been collected, the protocol can be followed, beginning with Subheading 3.1.2.

5. The timing of this step is not critical; the relevant endpoint is the dispersal of the cumulus cells from around the oocytes. There may still be a few adherent cumulus cells, this is not

critical as they will be removed with the zonae in subsequent steps.

6. Handling relatively small batches of oocytes makes the process more manageable as Acid Tyrode's solution will remove zonae very quickly (less than 30 s). The process can be slowed by increasing the pH to 3.0. Alternatively, the enzyme pronase (0.2%) can be used to digest the zonae pelucida.

7. The time required to dissolve zonae will vary among strains and should be monitored closely.

8. Prepare this solution under a fume hood due to the acute health effects of the beta-mercaptoethanol. The solution is good for 1 month.

9. Allow columns to come to RT before use.

10. To add DNase to the column, add one drop on the column. Let it soak then add another drop. Repeat the process until all 80 µl have been dispersed.

11. Good laboratory practices dictate that the preamplification work should be done in a different room than the rest of the protocol. The operator should not work on preamplified material and postamplified materials during the same work day. Use dedicated pipettes, equipment and even laboratory coat for preamplification and postamplification work. If it is not possible to do the preamplification work in a different room, use a bench as far away from the postamplification bench as possible. Limit foot traffic around the preamplification bench. Clean the benches frequently with diluted bleach.

12. The Klenow fragment of DNA polymerase I is lacking 5′ exonuclease activity.

13. The Illumina Paired-end adapter can be made as described below. We always make libraries using the paired-end adapters since these libraries can be sequenced on both single read and paired read flow cells. This ensures that one can sequence on a single read flow cell, then decide to sequence on a paired-end read flow cell without having to remake the library (the single end read adaptor are not compatible with the paired-end read flow cell).

   (a) Order the following two oligonucleotides from your preferred provider.

   • 5′ P-GATCGGAAGAGCGGTTCAGCAGGAATG CCGAG 3′

   • 5′ ACACTCTTTCCCTACACGACGCTCTTCCGA TC*T 3′

   *Indicates a phosphorothioate modification that protects the last nucleotide from exonuclease activity.

(b) Both oligonucleotides should be HPLC purified.

(c) To make the adapters, add 50 µl of both oligos at 100 µM each (oligos should be dissolved in STE buffer – 10 mM Tris-HCl, pH 8.0, 50 mM NaCl, 1 mM EDTA, pH 8.0). Denature at 95°C for 4 min, then transfer to 65°C and immediately add 1 µl 100 mM $MgCl_2$. Incubate at 65°C for 3 min, then place tubes at room temperature for 1 h. Dilute 1:10 in nuclease-free water and distribute to 0.2 ml tubes in a volume of 10 µl.

To quality control the adaptor, use an Agilent bioanalyzer with a DNA 1000 chip. Load the chip with the single-stranded oligonucleotides and the double-stranded adapter. The adapter lane should be free of single-stranded oligo-nucleotides, and display a unique band around 70 bp.

(d) We have also used custom adapters that have seven bases tags to create pool of libraries that can then be sequenced into the same sequencing lane to decrease cost. If you are using an Illumina GAIIx sequencer, keep the cluster density at 40 million per tile or less. You should always use a minimum of four bar coded adaptors per lane, and care must be taken to balance base composition as much as possible.

(e) This is a set of seven bases tag we have successfully used.

CACTGCC – CACTGCC – ACTTAGG – TAGCCTT – GTCGTTT – CAAGAGA – AAGACCT – TTCTGTG

These tags and their reverse complements must be added to both oligonucleotides to maintain the 3′ double-stranded structure of the adapters, and followed by a T to allow for the ligation to the acDNA with the A overhang. For example, if we use the first tag (CACTGCC), the oligonucleotide sequences in Note13a become:

P - G G C A G T G A G A T C G G A A G A G C G G T T C A G CAGGAATGCCGAG

A C A C T C T T T C C C T A C A C G A C G C T C T T C CGATCTCACTGCC*T

The newest image analysis software for the Illumina GAIIx (SCS 2.8, RTA 1.8) has difficulty with low complexity tags such as the ones above. If you have many raw clusters on the flow cell, the quality scores will drop dramatically and the software will flag most of the clusters as low quality. You will need to decrease the amount of DNA loaded on the flow cells. We use a 9 pM solution of denatured DNA and retrieve about 32–33 millions sequences after filtering (compared to 40 million for nonmultiplexed libraries using an 11 pM solution). Another strategy is to modify the RTA software setting so that it uses different bases to define clusters (the bases after your tags; in our example, bases 9–13). This strategy will only work if you have the T7500 Dell computer with your GAIIx. You can alter/add the following parameters in the C:\Illumina\ SCS2.X\RTA\RTA.exe.config file:

<add key = "FirstTemplateCycle" value="1">

<add key – "TemplateCycleCount" value="5">

If these two entries do not exist, just add them and alter the values for the desired cycles. If one still wants to use 5 cycles, instruct the RTA to set the First Template Cycle to 9, and keep the Template Cycle Count to 5. This will result in the system holding onto only 13–14 cycles worth of images; however, it will allow one to get past the bar codes before doing the cluster calling.

14. MinElute columns must be stored at 4°C but should be used at room temperature for the purification.

15. Try to minimize the amount of time the UV lamp is on. This will limit the damage to the DNA fragments. With the UV on, position a ruler on the gel to visualize where you want to cut the gel, turn the UV off and cut the gel.

References

1. Bertone P, Stolc V, Royce TE, Rozowsky JS, Urban AE, Zhu X, Rinn JL, Tongprasit W, Samanta M, Weissman S, Gerstein M, Snyder M (2004) Global identification of human transcribed sequences with genome tiling arrays. Science 306:2242–2246.

2. Schadt EE, Edwards SW, GuhaThakurta D, Holder D, Ying L, Svetnik V, Leonardson A, Hart KW, Russell A, Li G, Cavet G, Castle J, McDonagh P, Kan Z, Chen R, Kasarskis A, Margarint M, Caceres RM, Johnson JM, Armour CD, Garrett-Engele PW, Tsinoremas NF, Shoemaker DD (2004) A comprehensive transcript index of the human genome generated using microarrays and computational approaches. Genome Biol 5:R73.

3. Wilhelm BT, Marguerat S, Watt S, Schubert F, Wood V, Goodhead I, Penkett CJ, Rogers J, Bahler J (2008) Dynamic repertoire of a eukaryotic transcriptome surveyed at single-nucleotide resolution. Nature 453:1239–1243.

4. Nikolaev SI, Deutsch S, Genolet R, Borel C, Parand L, Ucla C, Schütz F, Duriaux Sail, G, Dupré Y, Jaquier-Gubler P, Araud T, Conne B, Descombes P, Vassalli JD, Curran J, Antonarakis SE (2009) Transcriptional and post-transcrip-

tional profile of human chromosome 21. Genome Research 19:1471–1479.

5. Robertson G, Schein J, Chiu R, Corbett R, Field M, Jackman SD, Mungall K, Lee S, Okada HM, Qian JQ, Griffith M, Raymond A, Thiessen N, Cezard T, Butterfield YS, Newsome R, Chan SK, She R, Varhol R, Kamoh B, Prabhu AL, Tam A, Zhao Y, Moore RA, Hirst M, Marra MA, Jones SJM, Hoodless PA, Birol I (2010) De novo assembly and analysis of RNA-seq data. Nat Meth 7:909–912.

6. Trapnell C, Williams BA, Pertea G, Mortazavi A, Kwan G, van Baren MJ, Salzberg

7. SL, Wold BJ, Pachter L (2010) Transcript assembly and quantification by RNA-Seq reveals unannotated transcripts and isoform switching during cell differentiation. Nat Biotech 28:511–515.

8. Guttman M, Garber M, Levin JZ, Donaghey J, Robinson J, Adiconis X, Fan L, Koziol MJ, Gnirke A, Nusbaum C, Rinn JL, Lander ES, Regev A (2010) Ab initio reconstruction of cell type-specific transcriptomes in mouse reveals the conserved multi-exonic structure of lincRNAs. Nat Biotech 28:503–510.

9. Haas BJ, Zody MC (2010) Advancing RNA-Seq analysis. Nat Biotech 28:421–423.

10. Trapnell C, Salzberg SL (2009) How to map billions of short reads onto genomes. Nat Biotech 27:455–457.

11. Marioni JC, Mason CE, Mane SM, Stephens M, Gilad Y (2008) RNA-seq: An assessment of technical reproducibility and comparison with gene expression arrays. Genome Res. 18(9):1509–1517.

INDEX

Wai-Yee Chan and Le Ann Blomberg (eds.), *Germline Development: Methods and Protocols*, Methods in Molecular Biology,
vol. 825, DOI 10.1007/978-1-61779-436-0, © Springer Science+Business Media, LLC 2012